Triassic-Jurassic Of Western Massachusetts

Easily Accessible Geology Field Trips

John F. Hubert

Front cover –

Cross-section of the Deerfield basin. Courtesy of Planetward.org.

View from Sugarloaf Mountain. Courtesy of John Burk.

Grooves and scour marks on the base of a river-channel sandstone in the Sugarloaf Arkose. The curved ends of the scours are up-flow, showing that the current was into the outcrop.

River channels in the Sugarloaf Arkose.

Cobbles and pebbles in the Sugarloaf Arkose. Courtesy of Kurt Hollocher.

Pillow structures in the Deerfield Basalt. Courtesy of the Massachusetts Geological Survey.

Hitchcock's dinosaur footprint quarry. Barton Cove, Gill.

Back cover -

As a handsomely dressed geologist, I lead a field trip to Connecticut in 1978. But why a plastic pocket protector when the pants have an ink stain and hole? The belt has a red paint stain.

A *Eubrontes* trackway is parallel to the crests of the lacustrine ripple marks, implying that the dinosaur was walking along the lake shore. Dinosaur Footprints Reservation, Holyoke.

Early morning beside a Triassic lake in the Hartford rift basin. The 21-foot long phytosaur *Rutiodon* snatches a *Semionotus* from the shallows. The dinosaur maker of *Eubrontes* footprints passed this way the previous evening. My daughter Ananda made the drawing by scratching away a black coating to reveal the white underneath.

Second edition. Copyright 2018 by John F. Hubert.

ISBN-10:1977733646

ISBN-13: 978-1977733641

All rights reserved.

For Mary Alice Gorman Hubert (1929-2012), my wife of 56 years who warmed her hands on the fire of life. Thank you.

Contents

Preface	ix
Introduction	1
Hartford Basin	1
Deerfield Basin	10
Field trips in the Hartford basin	12
1. Summit of Mount Tom, Holyoke	12
2. Dinosaur Footprints Reservation, Holyoke	19
3. Playa and lacustrine strata, Holyoke Community College	28
4. Dinosaur Park and Arboretum, Rocky Hill	35
Field trip in the Amherst block	42
5. Beneski Natural History Museum, Amherst College	42
Field trips in the Deerfield basin	48
6. Summit of Sugarloaf Mountain, South Deerfield	48
7. Fluvial redbeds, Route 116 at Sugarloaf Mountain, South Deerfield	62
8. Poet's Seat Tower, Greenfield	70
9. Fluvial redbeds, Country Club Road, Greenfield	75

10. Alluvial fan, West Gill Road, Gill 85

11. Lacustrine strata, Barton Cove Campground, Gill 89

12. Lacustrine delta and alluvial fan, Chard Pond, Sunderland 102

13. Mount Toby Conglomerate, Roaring Brook, Sunderland 116

14. Deerfield Basalt, Route 2, Gill 128

15. Sugarloaf Arkose and hydrothermal hot spot, Greenfield 138

16. Geo Path and Rock Park, Greenfield Community College 147

References Cited 152

Web Credits 156

Glossary 162

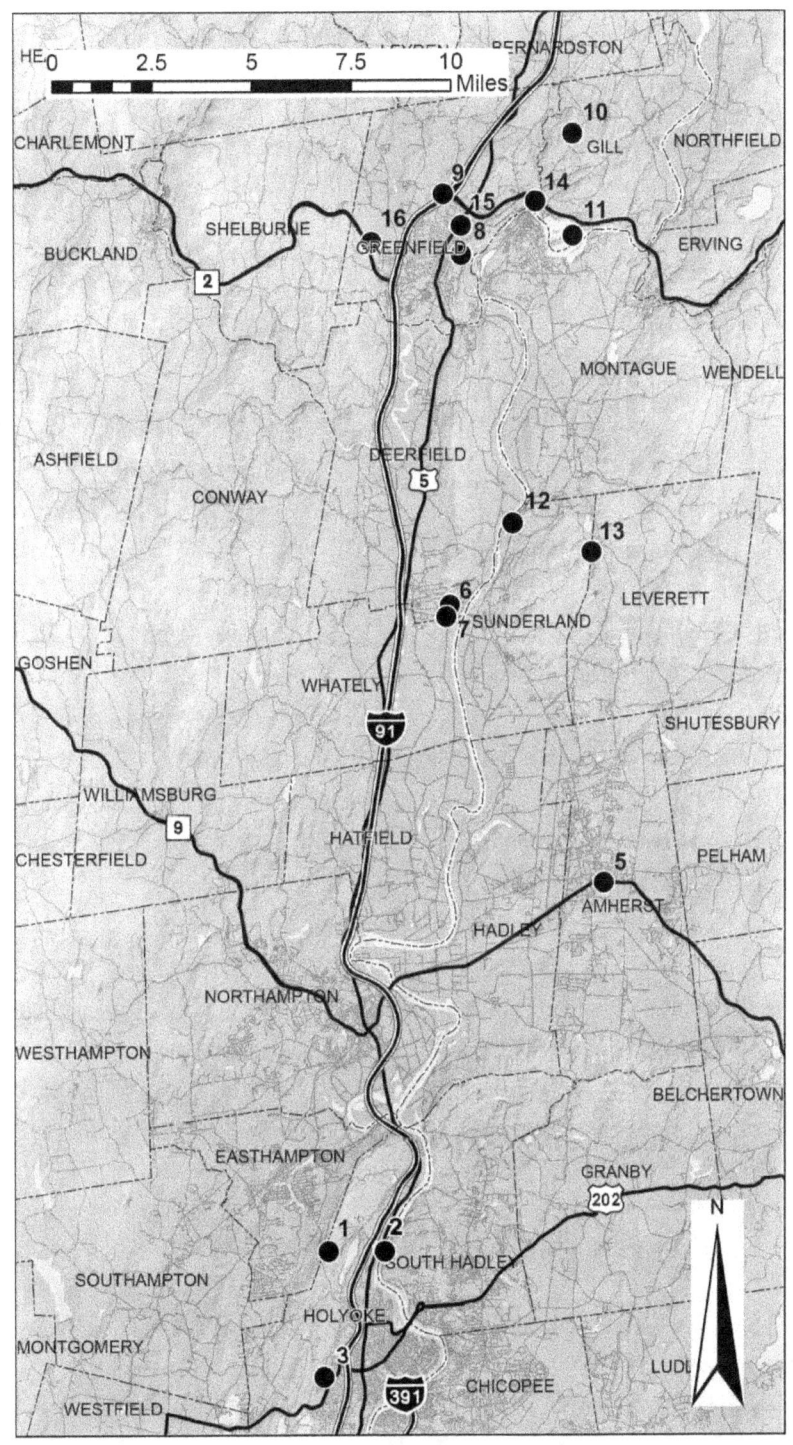

Location of the field trips in western Massachusetts. Not shown is trip 4 to the Dinosaur Park and Arboretum in Rocky Hill, Connecticut, a few miles south of Hartford.

	Location		Latitude	Longitude
1	Summit of Mount Tom	Holyoke, MA	42.2415	-72.6482
2	Dinosaur Footprints Reservation	Holyoke, MA	42.2424	-72.6229
3	Playa and lacustrine strata, HCC	Holyoke, MA	42.1964	-72.648
4	Dinosaur Park and Arboretum	Rocky Hill, CT	41.6519	-72.6568
5	Beneski Museum, Amherst College	Amherst, MA	42.3723	-72.5143
6	Summit of Sugarloaf Mountain	S. Deerfield, MA	42.4702	-72.5921
7	Fluvial redbeds, Route 116 at Sugarloaf Mountain	S. Deerfield, MA	42.4676	-72.5929
8	Poet's Seat Tower	Greenfield, MA	42.5955	-72.5867
9	Fluvial redbeds, Country Club Road	Greenfield, MA	42.6161	-72.5978
10	Alluvial fan, West Gill Road	Gill, MA	42.638	-72.5335
11	Lacustrine strata, Barton Cove Campground	Gill, MA	42.6034	-72.5327
12	Lacustrine delta and alluvial fan, Chard Pond	Sunderland, MA	42.5	-72.5604
13	Mount Toby Conglomerate, Roaring Brook	Sunderland, MA	42.4895	-72.5234
14	Deerfield Basalt, Route 2	Gill, MA	42.6133	-72.5525
15	Sugarloaf Arkose and hydrothermal hot spot	Greenfield, MA	42.6057	-72.5888
16	Rock Path and Rock Park, GCC	Greenfield, MA	42.5997	-72.6325

Preface

These field trips are a hands-on introduction to evidence of the landscapes and life during the late Triassic to early Jurassic in western Massachusetts. The excursions are in the Deerfield and Hartford basins, which comprise the Connecticut Valley basin, plus the Beneski Natural History Museum of Amherst College located between the two basins. Some excellent outcrops are omitted because stopping on divided highways is prohibited by law.

The trips do not require strenuous effort, except possibly for the Mount Toby Conglomerate at Roaring Brook in Sunderland/Leverett. Wear comfortable boots because it is easy to turn an ankle with sneakers. I recommend you carry a 10X hand lens to observe details of the rocks.

To be sure of the route when planning a trip, check on Google Earth and Google maps.

The trips are intended to require no previous geologic training. I have tried to define geologic jargon, or make the meaning clear by the context, but if unclear, the web is usually helpful.

Both metric and English scales are used, in part because many of the diagrams have metric units embedded in them.

I thank Jim Dutcher, long-time friend and co-researcher on Triassic-Jurassic redbeds, for his help with this project. Jim took many of the photographs while we were together in the field. Jim introduced his students at the Holyoke Community College to the outcrop on campus and suggested the site for this guide.

Nick Venti prepared the location map of the field trips and the table of values of latitude and longitude.

The volume is only possible because of the efforts of my graduate and undergraduate students who taught me more than I ever taught them. The insights of Don Wise clarified structural and tectonic concepts of the rift basins.

All illustrations are the author's, in the public domain, from open-source sites such as Wikipedia and the U.S. government, under the fair-use or author's rights provisions of copyright law, or the copyright holder is acknowledged under the illustration and in the web or book credits.

This guide is possible because of the help of Mr. Wikipedia and Ms. Google who seem to know everything. Don Sluter and Nick Venti got me get a little closer to the twenty-first century in computer matters, including the delightfully named Drop Box in the cloud. For decades, I used a slide rule, typewriter, Kodachrome slides, and India-ink line drawings

There is truth in the old joke:

If you steal from one author, it's plagiarism;

If you steal from two or three authors, it's literary discernment;

If you steal from many, it's masterful research.

Please inform me of the errors and omissions so I can correct them.

Introduction

Beginning 230-million years ago in the late Triassic, the Deerfield and Hartford rift basins were created by regional N70W-S70E extensional pull-apart associated with doming of the Earth's crust during the early stages of break up the supercontinent Pangea. The basins are part of the Newark Supergroup along eastern North America that was a failed attempt to open the Atlantic Ocean, which happened later starting about 180 million years ago. More than five kilometers of terrestrial strata and basalt lavas accumulated in the basins as they drifted north in the subtropics at about 25°N paleolatitude. After the initial stage of basin sag, border faults were active on the east sides of the basins. Integration of the border faults into the present continuous border fault happened later during the Cretaceous.

The geological time chart shows that the Earth began 4.6 billion years ago. The Precambrian is one word, but it represents more than 80 percent of the time. If the 4.6 billion years is modeled by the 102 stories of the Empire State Building in New York City, then human time on Earth is less than the thickness of a dime placed on the top of the building.

Hartford Basin

The Hartford basin is 80 miles long, from New Haven to Northampton. The maximum width of 25 miles is at the Connecticut/Massachusetts border. The 4.5 km of preserved alluvial-fan, fluvial, and playa redbeds, lacustrine gray/black strata, and intrusive and extrusive basalts, were deposited during 50 million years in the late Triassic-early Jurassic. Although playa is Spanish for a beach, in geomorphology a playa is a desert basin with no outlet that during a storm partially fills with water to form a temporary lake.

In the initial stage of basin sag in the late Triassic, rivers flowed from both sides of the basin to deposit the lower part of the 2-km-thick New Haven Arkose. An arkose is a feldspar-rich sandstone. The feldspar group includes aluminum silicates of sodium, potassium, or calcium and is the most abundant mineral group on Earth. The upper part of the New Haven Arkose formed while a border fault was active on the east side of the rift valley with the sand and gravel coming from mountainous terrane east of the fault. The subsidence of the rift valley floor provided the accommodation space to allow accumulation of vertical kilometers of sediment.

U.S. Geological Survey Time Chart
Units in millions of years

Quaternary	2.5 – present	Holocene	0.01 – present
		Pleistocene	2.5 – 0.01
Cenozoic	66 – 1.8	Pliocene	5.3 – 1.8
		Miocene	23 – 5.3
		Oligocene	34 – 23
		Eocene	56 – 34
		Paleocene	66 – 56
Mesozoic	251-66	Cretaceous	146 – 66
		Jurassic	200 – 146
		Triassic	251- 200
Paleozoic	542 – 251	Permian	300 – 251
		Pennsylvanian	318 – 300
		Mississippian	359 – 318
		Devonian	416 – 359
		Silurian	444 – 416
		Ordovician	488 – 444
		Cambrian	542 – 488
Precambrian	4600 – 542	Proterozoic	2500 – 542
		Archean	4000 – 2500
		Hadean	4600 – 4000

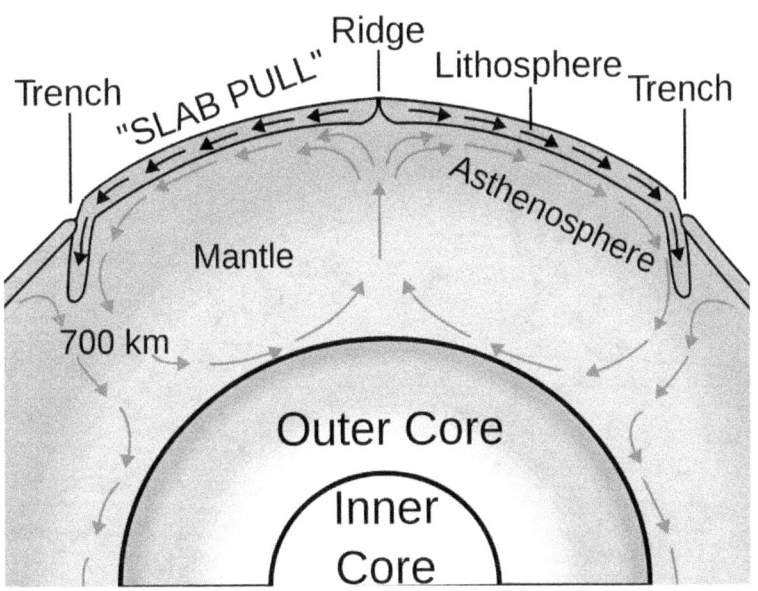

A tectonic plate is a piece of the crust and upper mantle that moves over the asthenosphere, a relatively soft, plastic layer located 80 to 200 km below the surface. Courtesy of age-of-the-sag.org.

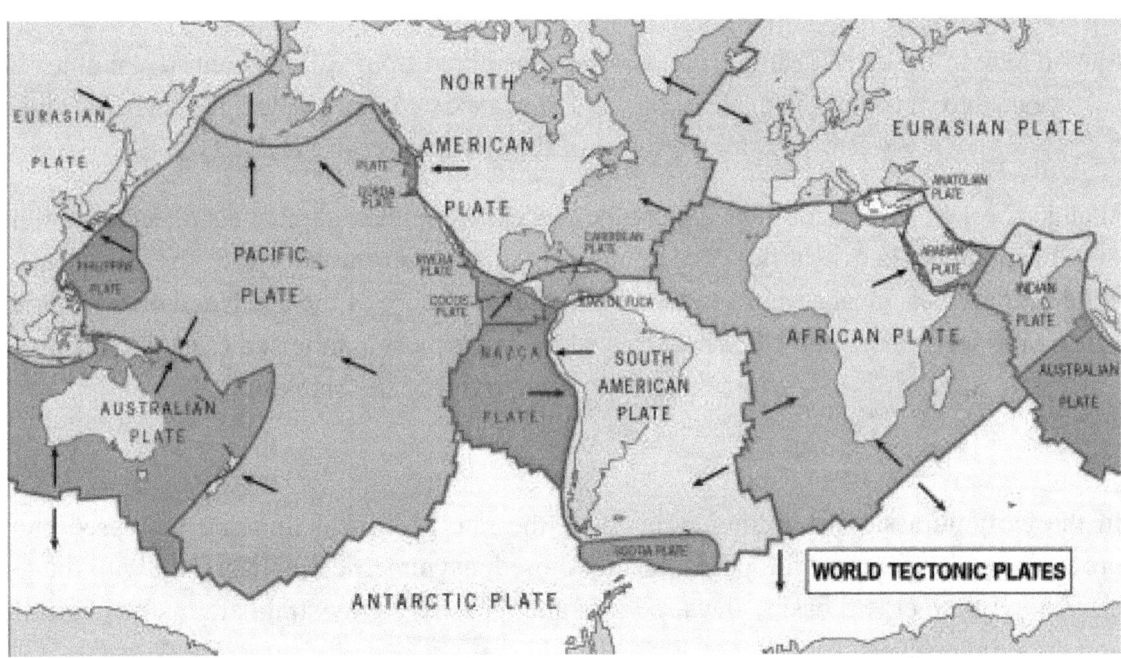

The arrows show the direction of movement of the Earth's tectonic plates. The plates are always in motion, sliding past, into, under, and over each other. Convection cells in the mantle drive plate motion at the rate your fingernails grow. Courtesy of Yale Scientific magazine.

Shown here in the early Triassic, Pangea was assembled from various continental units 335 million years ago. Each unit is made of plates. Pangea existed during the late Paleozoic and early Mesozoic eras. Courtesy of the Gloucestershire Geology Trust.

Pangea broke apart because the assembled pieces of continental crust act as an insulation blanket over the mantle where heat flow is continuously moving upwards towards the crust. The sources of the heat are heat emanating from the hot, liquid outer core of the Earth and decay of radioactive elements, such as uranium. The heat caused the uppermost mantle and crust to expand, forming a dome that over time produced the breakup.

In the early Jurassic 205 million years ago, the rate of crustal thinning increased substantially, producing volcanism and a topographically closed, asymmetrical half-graben with the fault on the east side. In the closed basin, playa redbeds and lacustrine gray strata alternated controlled by wet and dry climate cycles.

Rivers and sheet floods from the western highlands spread sands and muds over the playas, which extended up to small alluvial fans along the fault-bounded low cliff of the eastern escarpment. These playa-lacustrine-fluvial strata make up the 150-m thick Shuttle Meadow, 170-m-thick East

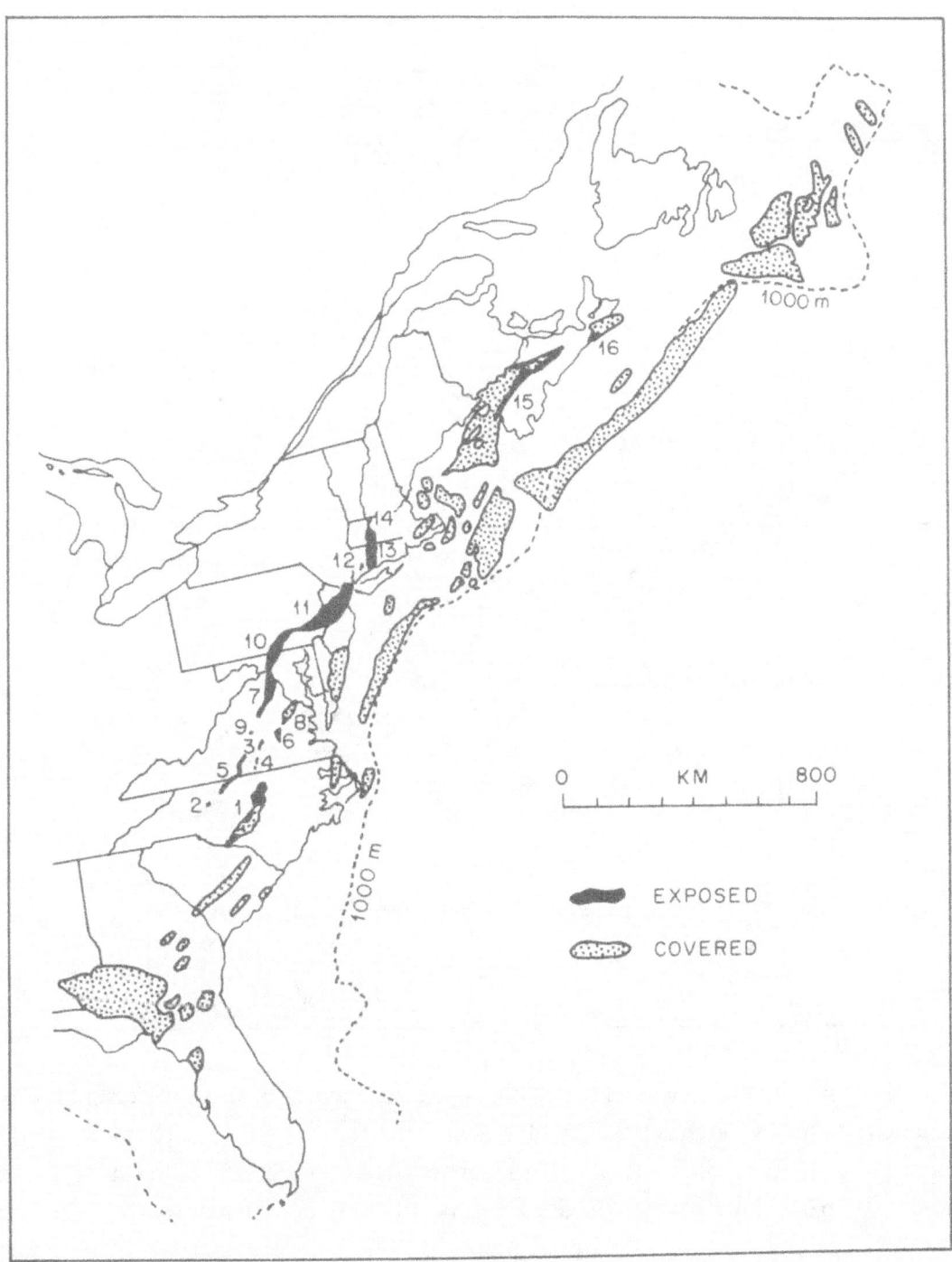

The Newark Supergroup of Triassic-Jurassic basins formed during the initial attempt to break up Pangea and open the North Atlantic. The Hartford (13) and Deerfield (14) basins are part of the 16 basins exposed on land. The covered basins lack surface exposure of the rocks, but are larger and more numerous than the exposed basins. Courtesy of Van Houten.

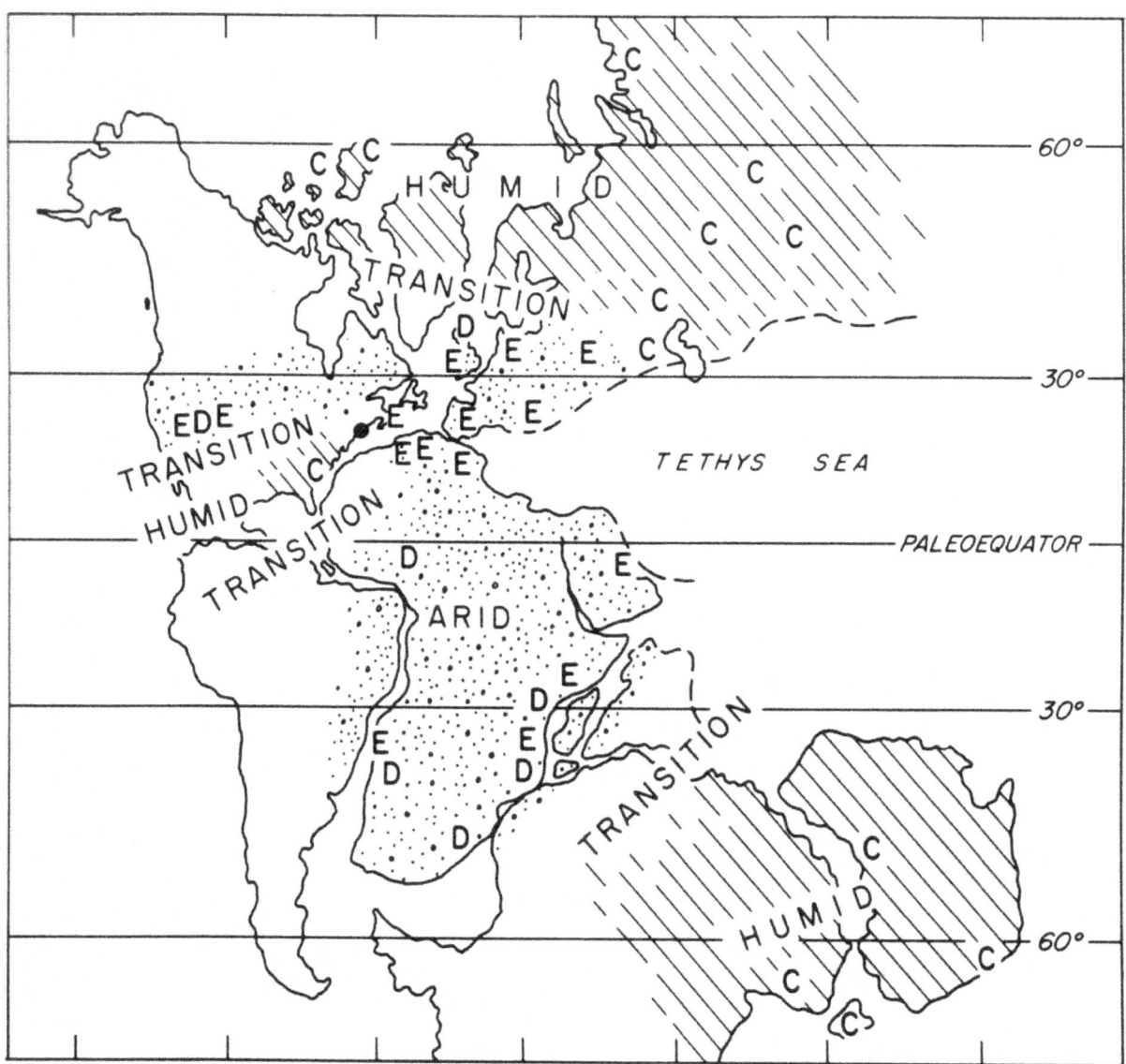

The Late Triassic climate zones in Pangea are based on climate-sensitive rocks: coal (C), dune sandstone (D), and evaporates (E). The black dot is the Hartford and Deerfield rift basins. In response to continental drift, the rift valleys moved slowly northward in the subtropics in a seasonal wet/dry paleoclimate where the dry season commonly dominated over the wet. Modified from Robinson, 1973.

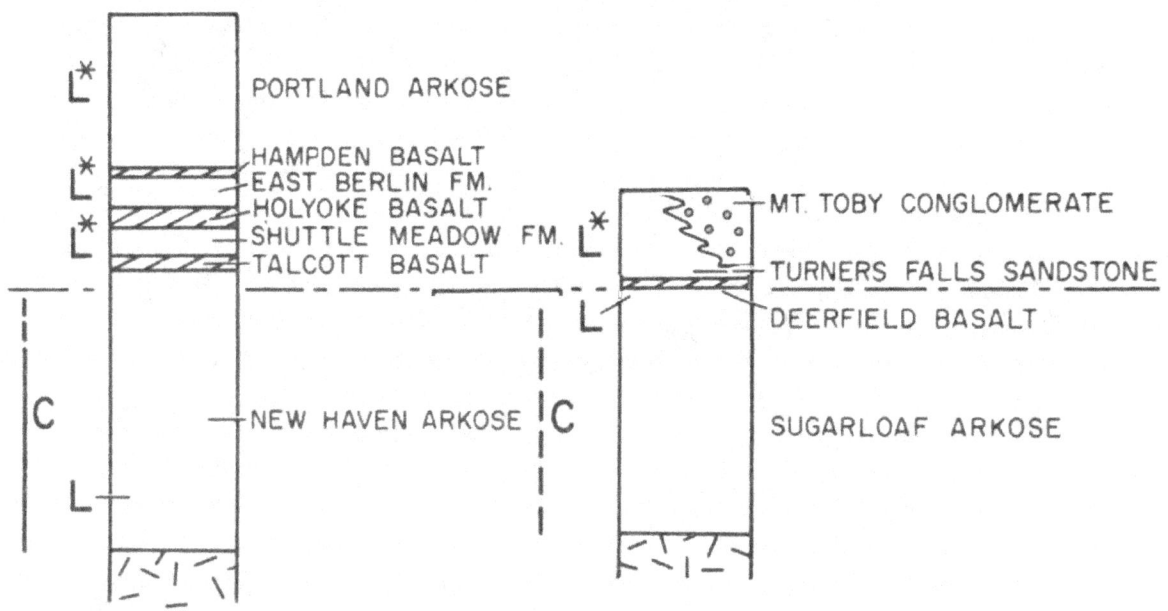

The Triassic-Jurassic boundary in the Hartford and Deerfield basins is now placed a bit lower in the column. The thin stratigraphic unit of the Falls River beds has been introduced below the Deerfield Basalt. Caliche is a calcareous soil that forms in semiarid to arid climates. Analcime is a sodium-rich silicate mineral precipitated by evaporation of shallow lakes. From Hubert *et al.*, 1978.

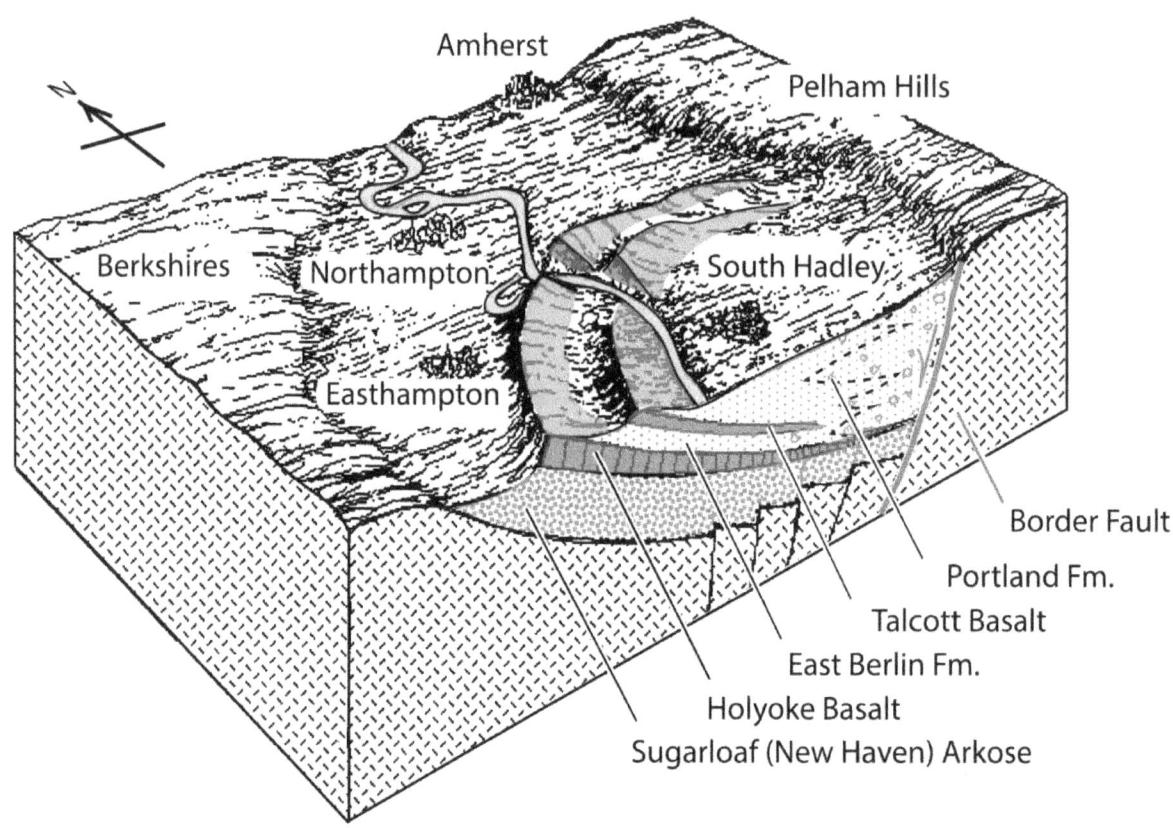

At the north end of the Hartford basin just south of its junction with the Deerfield basin, the Holyoke Range becomes the Mount Tom Range where the east-west trend changes to a more southerly direction. As deposited, the Hartford and Deerfield basins comprised a single basin. For example, the Holyoke and Deerfield basalts were a continuous volcanic unit, now separated by erosion at the uplifted block of lower Paleozoic rocks near Amherst. The basalts have different names because the basins were first studied in the nineteenth century before the tectonic history was known. In the diagram, the stratigraphic names apply to the Hartford basin. The dashed lines in the Portland Formation indicate the presence of alluvial fans. Marie Litterer drafted the figure. Courtesy of Don Wise.

Berlin, and lower half of the 2-km-thick Portland formations. A geological formation is a mappable unit of strata, named for the geographic area where first described.

Lavas fed by magma eruptions from deep-seated fissures produced three lava-flow units: the 65-m-thick Talcott basalt that separates the New Haven Arkose from the Shuttle Meadow Formation;

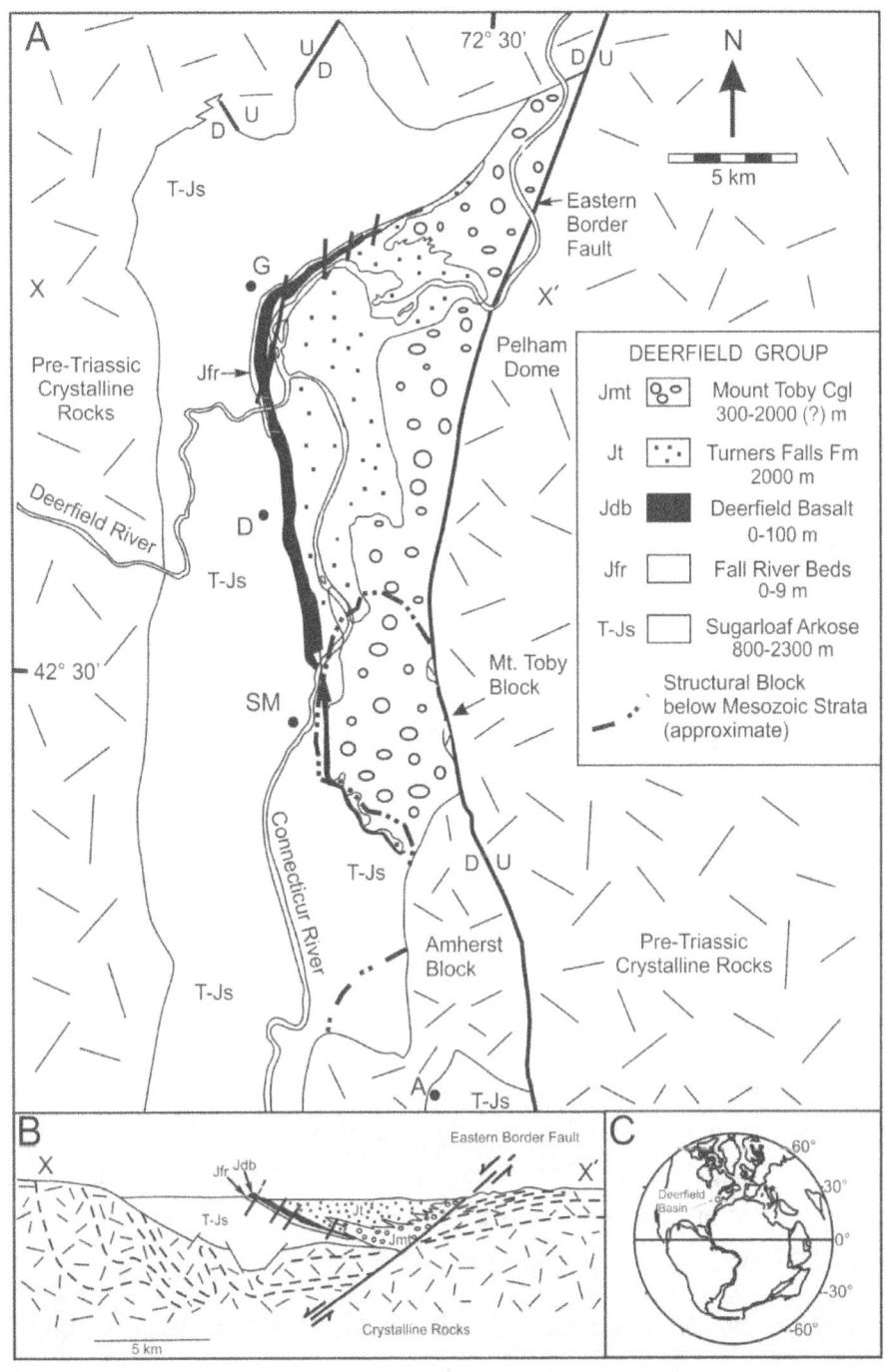

The strata in the Deerfield basin were deposited in horizontal orientation and later folded into a syncline tilted to dip towards the border fault. Place names are Sugarloaf Mountain (SM), Amherst (A), Greenfield (G), and Deerfield. Insert C is Pangaea. The map is from Olsen *et al.*, 1992. The cross section is adapted from Zen, 1983.

the 100-m Holyoke basalt between the Shuttle Meadow and East Berlin formations; and the 60-m Hampden basalt between the East Berlin and the Portland formations.

As extension eased, the fault-bounded eastern highlands shed detritus for the fluvial and alluvial-fan redbeds of the upper Portland Formation.

The dinosaur footprints and tracks of the Connecticut Valley make their first appearance in the playa and lacustrine sandstones and mudstones of the Shuttle Meadow Formation. Similar impressions are present in the same type of strata in the East Berlin and Portland formations. Although dinosaurs lived along the rivers, the fluvial environment is not favorable for the preservation of tracks.

Deerfield Basin

The Deerfield basin is 25 miles long from the Holyoke River to Bernardson and 7 miles wide. The basin has over 4 km of preserved sedimentary rocks and the Deerfield Basalt lava flows.

The Sugarloaf Arkose was the first stratigraphic unit deposited during the initial pull-apart extensional sag of the basin. The formation comprises 800 to 2,300 meters of fluvial redbeds, recording about 16 million years during the latest Triassic to earliest Jurassic. The upper surface of the Sugarloaf Arkose is a slight angular unconformity that formed in response to rapid subsidence and tilting of the basin floor along the eastern border fault. The beds over the arkose are the three to 9 meters of the Fall River beds that accumulated in marsh, lake, and playa settings.

Over the Fall River beds is the 0 to 100-meter-thick Deerfield Basalt, consisting of two flows. The Deerfield and Holyoke basalts are each made of the same two flows, pouring out as fissure eruptions that covered much of southern New England.

After the volcanic activity, rapid subsidence maintained a topographically-closed basin in which accumulated the 2,000 meters of the Turners Falls Formation, a sequence of playa and fluvial redbeds interbedded with gray/black lacustrine strata. The basin was located between humid conditions to the south and aridity to the north, so that the climate shifted from wetter to drier and back again, favoring formation of lakes during the wetter phase.

During accumulation of the Turners Falls Formation, alluvial fans built the 300 to 2,000 meters of the Mount Toby Conglomerate along the escarpment of the eastern border fault. The conglomerates are in part laterally equivalent to the playa and lacustrine deposits of the Turners Falls Formation. As in the Hartford basin, dinosaur and other tracks are present in some of the playa and lacustrine beds of the Turners Fall Formation.

Downward movement on the border fault folded the strata into a syncline tilted towards the border fault.

Field Trips in the Hartford Basin

1. Summit of Mount Tom, Holyoke

What to see. - From its perch 1,201 feet above the Connecticut River, the summit of Mount Tom provides a view of the valley, the Berkshire Mountains to the west, and Pelham hills to the east. You can see the oxbow bend in the Connecticut River, columnar jointing (palisades) in the early Jurassic Holyoke basalt, hike the trails, and visit the site of the former Mount Tom Hotel, a popular destination in the early 1900s.

Most of Mount Tom is in the Mount Tom State Reservation of 2,161 acres with 22 miles of walking trails, picnic facilities, fishing in Lake Bray, and in the winter cross-country skiing and ice skating. In September, the mountain is a popular place to watch migrating hawks and falcons fly by, commonly at elevations below the summit peak. Some days hundreds of birds can be seen. The migrators include the Red-tailed Hawk, Cooper's Hawk, American Kestrel, Peregrine Falcon, and Northern Harrier.

How to get there. - The Massachusetts Office of Energy and Environmental Affairs gives these directions to get to Mount Tom. The daily parking fee is five dollars for a Massachusetts vehicle and six if out-of-state.

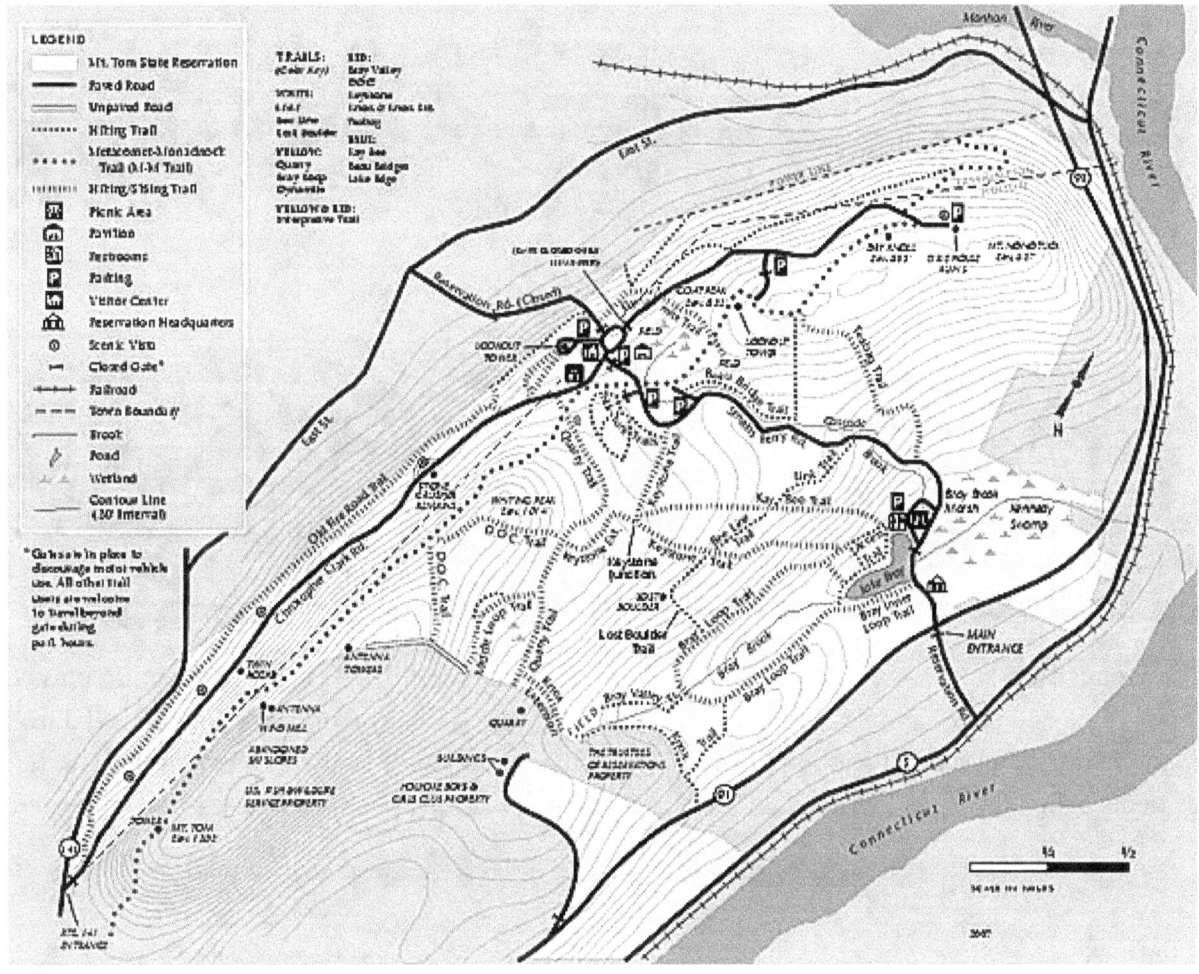

Trail map of Mount Tom State Reservation. Courtesy of http://www.mappery.com.

From the west:

1. Head east on I-90 E.
2. Take exit 4 for I-91 toward Holyoke/Springfield - 0.6 mi.
3. Keep left at the fork, follow signs for I-91 and merge onto I-91 N - 3.9 mi.
4. Take exit 17A to merge onto Easthampton Rd toward Holyoke - 0.5 mi.
5. Turn left at Northampton St/US-5N - 4.0 mi.
6. Turn left at Reservation Rd. Park will be on the left 0.7 mi.

The Mount Tom Hotel was built in 1897 but burned down in 1900. The hotel was rebuilt, burned again in 1929, and never rebuilt. In 1897, the Holyoke Street Railway Company constructed a trolley park and later an amusement park on the east side of the mountain, which was open until 1988. Local residents called the park, "The Queen of the Mountain." Antique postcard courtesy of payle.com.

From the east:

1. Head west on I-90.
2. Take exit 4 for I-91 toward Springfield/Holyoke - 0.2 mi.
3. Keep left at the fork, follow signs for I-91 and merge onto I-91 N - 3.9 mi.
4. Take exit 17A to merge onto Easthampton Rd toward Holyoke - 0.5 mi.
5. Turn left at Northampton St/US-5N - 4.0 mi.
6. Turn left at Reservation Rd. Park will be on the left - 0.7 mi.

From the north:

1. Take I-91 S.
2. Take exit 18 to merge onto Mount Tom Rd/US-5S - Continue to follow US-5S - 3.8 mi.
3. Turn right at Reservation Road.
4. Park will be on the left - 0.7 mi.

Left: The Hockanum Bend before formation of the oxbow in 1840. Right: The oxbow in 1910. Professor Edward Hitchcock of Amherst College with his students would climb Mount Tom and lecture on the oxbow. Student tradition is that they were viewing the 1840 flood of March 3 to 4 when the raging Connecticut River cut through the narrow neck of the loop to create the oxbow, separating the bend from the river. I like to think that Professor Hitchcock said, "OK, that's how an oxbow forms." The Hockanum Bend image is by O. H. Troop published in 1826 in *The Northern Traveler*. Antique postcards courtesy of Wikipedia.

The boats in the oxbow marina can get on the Connecticut River by a narrow connection. The bend in the Holyoke-Mount Tom Range is in the upper right. Courtesy of dmampo.org.

The columns in the columnar jointing of the palisades at Mount Tom are bounded by fractures in the Holyoke Basalt that formed due to stress as the lava cooled because the solid basalt occupies less volume than the liquid. The polygonal columns have 3 to 8 sides, 6 being most common. Courtesy of Wikipedia.

A hang glider soars from the Holyoke Basalt cliff. The skyline is the plain formed by fluvial erosion in the Miocene, named a peneplain. Courtesy of Mount Holyoke Range State Park and Mass.gov.

From the south:

1. Take I-91 N.
2. Take exit 17A to merge onto Easthampton Rd toward Holyoke - 0.5 mi.
3. Turn left at Northampton St/US-5N - 4.0 mi.
4. Turn left at Reservation Rd. Park will be on the left - 0.7 mi.

Comments. – Tony Philpotts (2012) writes, "The lower Jurassic Holyoke basalt is one of the world's largest flood-basalt flows, with a volume in excess of 1200 km^3. It was fed by the 50 m-wide Buttress diabase dike, which can be traced through the crystalline basement rocks of Connecticut and Massachusetts on either side of the Hartford Basin and up through the Mesozoic sedimentary rocks of the basin, almost to the point of contact with the flow....It must have erupted rapidly, because it solidified as a single cooling unit, which is up to 200 m thick against the eastern border fault of the Mesozoic Hartford Basin....The immediate source of the magma is ... a mid-crustal reservoir at a depth of 12 km.... Magma would have risen to the surface from this depth through the 50 m-wide dike in only minutes....Upon eruption, the Holyoke basalt formed a thick lava lake (sea), which would have taken approximately 100 years to solidify."

A legend says that Mount Tom is named after Rowland Thomas who worked as a surveyor for the town of Springfield in the 1660s. He may have named the peak after himself when his fellow surveyor was working on the other side of the Connecticut River.

Fountains play in the world's largest lava lake located in the 2-km wide crater of volcanic Mount Nyiragongo in the Republic of the Congo. The lava lake is 600 m deep with a volume of about 1.9 km^3, not much compared to the Holyoke basalt lava lake "in excess of 1200 km^3" (Philpotts, 2012. Courtesy of Wikipedia.

2. Dinosaur Footprints Reservation, Holyoke

What to see. - You can walk on a bedding surface with numerous dinosaur footprints and trackways in playa redbeds in the lower part of the early Jurassic Portland Formation. Invertebrate burrows and ripple marks are also present.

How to get there. - From I-91, take exit 18 in Northampton and proceed south on Route 5 for about 4 miles. Park at the pull-off on the east side of Route 5 in Holyoke, not far from the Easthampton town line. The site is open daily sunrise to sunset from April 1 to November 30, but closed during the winter due to slippery conditions.

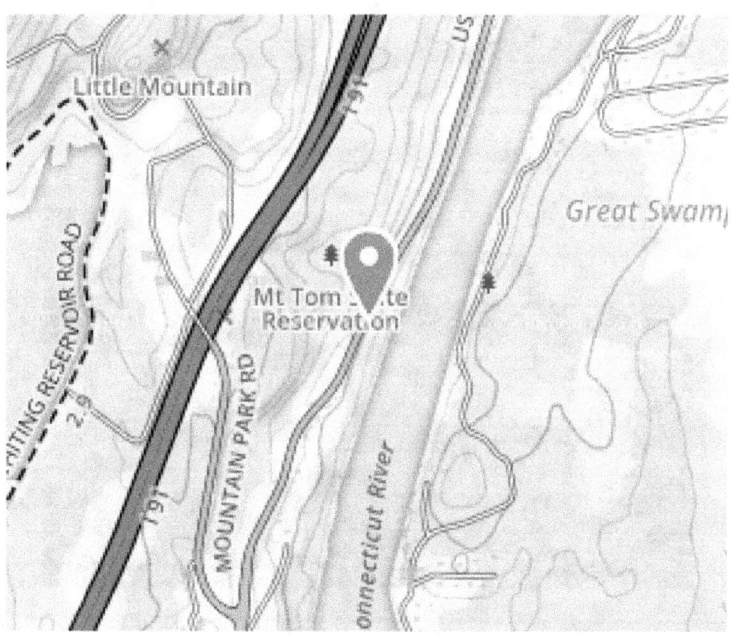

Location of Dinosaur Footprints Reservation adjacent to Route 5 in Holyoke. Courtesy of Google maps.

Pause at the information board, then proceed down the path to the bedding plane with tracks at the end of the path. The exposure is 40 m long and 4 to 8 m wide. Courtesy of http://explorewmass.blogspot.com.

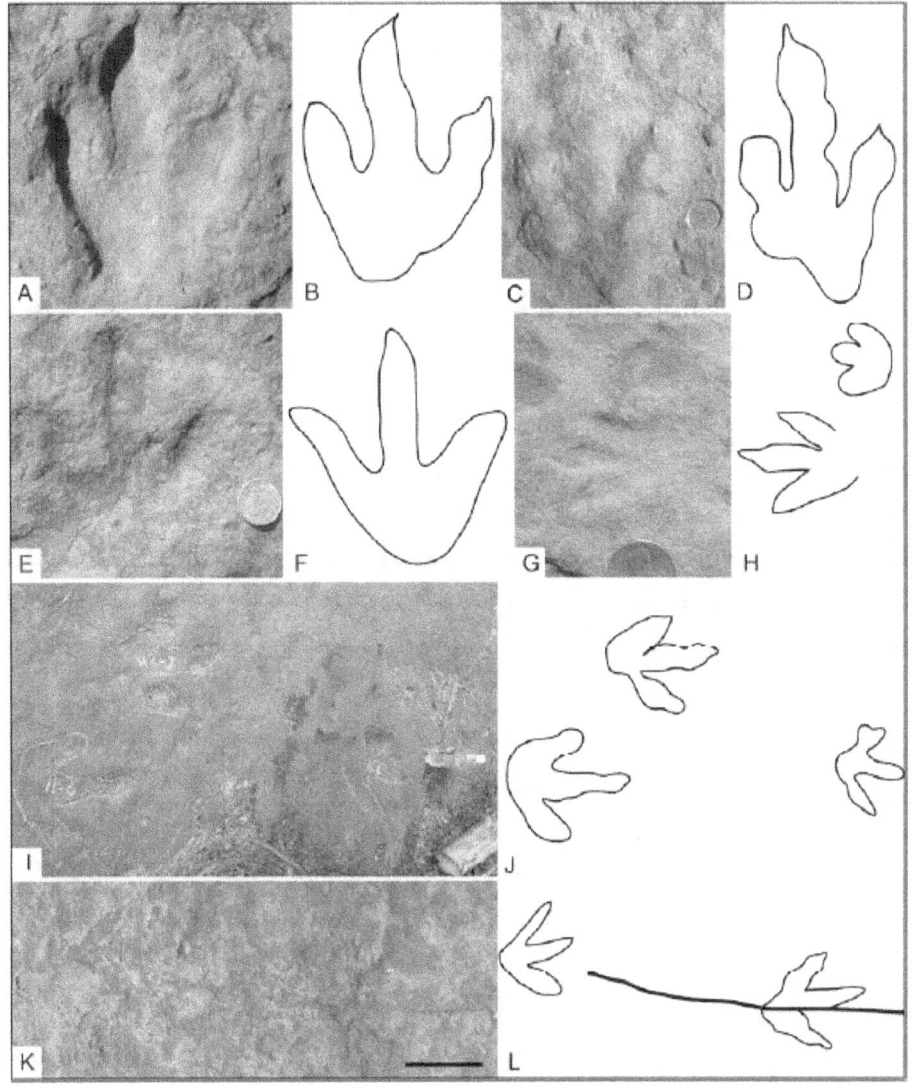

Vertebrate tracks preserved on the bedding surface. From Getty *et al.*, 2012.
A and B. *Eubrontes giganteus*.
C and D. *Anchisauripus*. Note the pads on the toes.
E and F. *Anomoepus scambus*, a small herbivorous ornithischian. Note the wide angle between the toes, stubby middle digit, and blunt claws that are characteristic of this ichnospecies.
G and H. Crocodylomorphic *Batrachopus*. The manus is smaller and at the top of the image; the pes is larger and at the bottom. The pes is missing the first digit.
I and J. Oblique view of the turning *Eubrontes* trackway.
K: Close-up of the *Anomoepus* trackway with a trail drag.
L: Interpretive drawing of K. Coin in A = 2.1 cm, in C and E = 2.4 cm, and in G = 1.9 cm. The bottle (right center) in I is approximately 20 cm tall, scale bar in K is 10 cm.
Courtesy of Paul Olsen and Peter LeTourneau.

The bedding surface with tracks is down a dirt path that leads to the north from the pullover. The site is maintained by the Massachusetts Trustees of Reservations, a member-supported, non-profit land conservation and historic preservation organization dedicated to preserving natural and historical places. Vandals can be fined up to $300.

Comments. - This site is well known to paleontologists because the late Professor John Ostrom of Yale University in 1972 wrote a paper where he said the dinosaurs travelled in a "herd, pack, or flock." Ostrom concluded that the large animals were traveling in a group because of the mostly east to west travel direction of the *Eubrontes giganteus* trackways, combined with the assumption of no physical barrier.

Ostrom assigned the tracks to small, medium, and large theropod dinosaurs, namely the ichnospecies *Grallator cuneatus*, whose prints are 76 to 127 mm long, *Anchisaurus sillimani* and *Eubrontes giganteus,* which are 279 to 330 mm long. In the 1830s, Professor Edward Hitchcock at Amherst College cited the quarry location as the source of his type specimen of *Eubrontes* ("true thunder") *giganteus.*

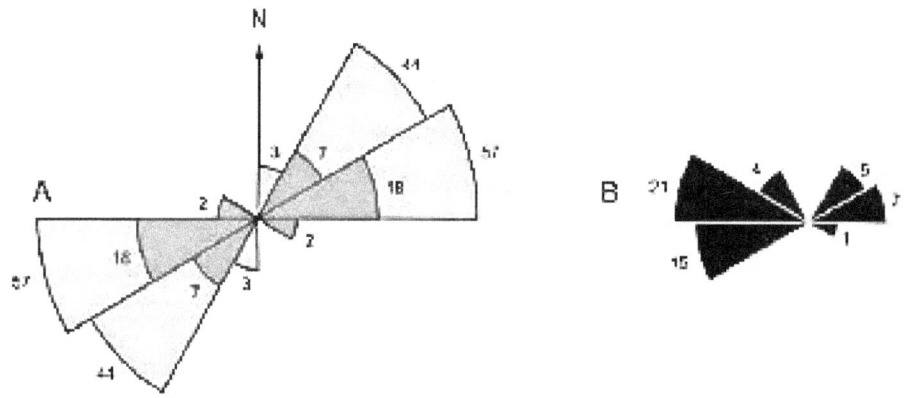

A) Ripple crests below (darker pattern) and above the main track horizon suggest a northeast-southwest trend to the lake shoreline. B) *Eubrontes* trackways on the main track bed. The trackways are approximately east or west, subparallel to the inferred shoreline. From Getty *et al.*, 2012.

For many years it has been known that the *Eubrontes giganteus* trackways are nearly parallel to the northeast-southwest orientation of the crests of wave-formed ripples, implying that the dinosaurs walked along the shore of a shallow lake, but not necessarily in a group. As wind-driven waves approach the shore of a lake, they refract to become parallel to the shore. However, some other dinosaurs were gregarious based on numerous sites around the world.

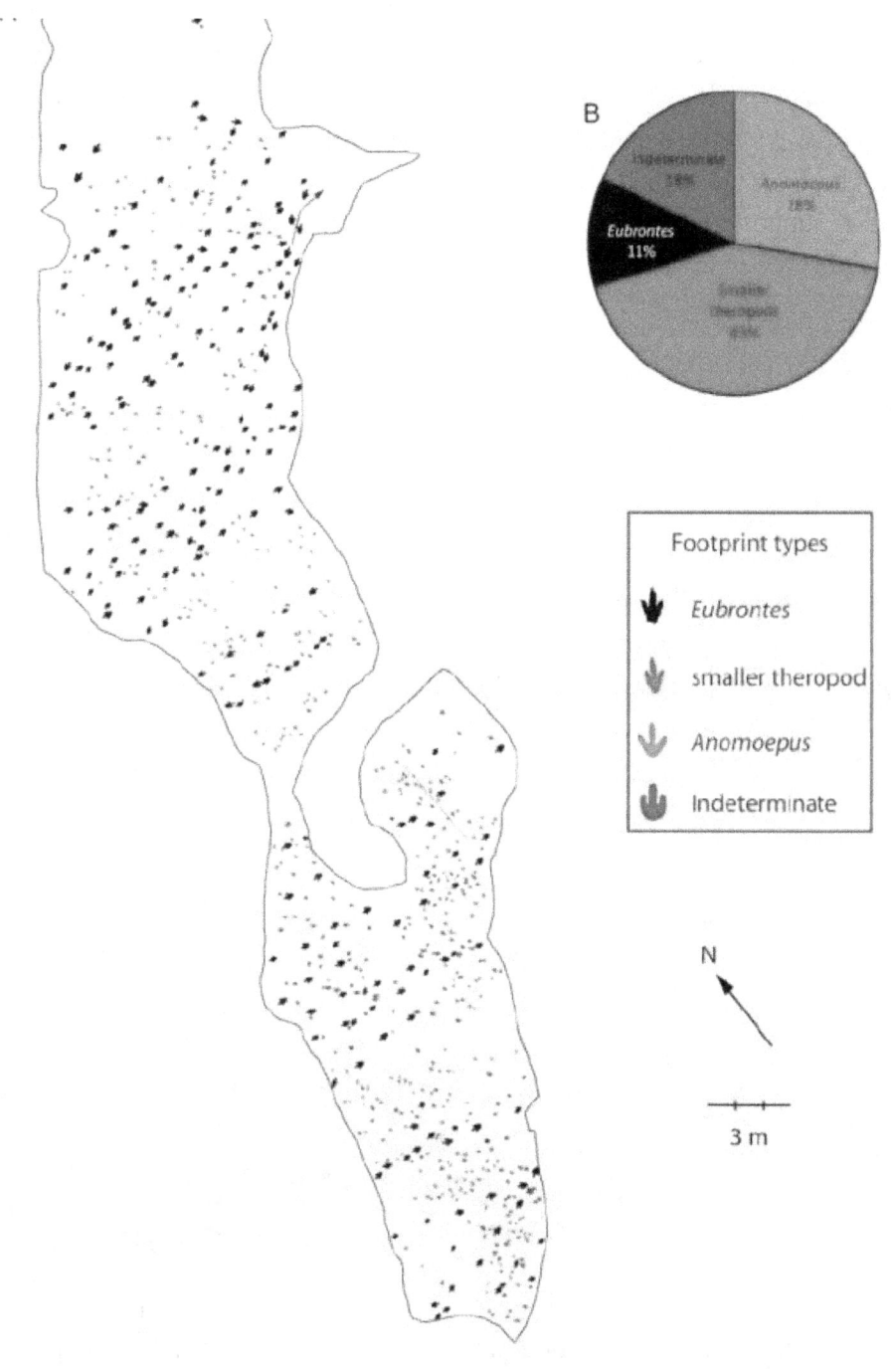

A map of the trackways shows that the critters were mostly waking to the northeast or southwest parallel to the inferred trend of the lakeshore. From Getty *et al.*, 2012.

Wikipedia notes: "The latest mapping project, conducted by Patrick Getty and Aaron Judge, has shown that there are at least 787 dinosaur tracks at the site and that the *Eubrontes giganteus* trackways are in fact parallel, or nearly parallel, to the orientation of oscillation wave-formed ripples. Considering that oscillation ripples form parallel to the shoreline, these authors suggested that the parallel trackways represent shoreline-paralleling behavior in large carnivores rather than group behavior. The hypothesis that the parallel trackways were made by shoreline-paralleling behavior is further supported by the fact that parallelism is not seen in *Eubrontes giganteus* trackways preserved at other sites in the Connecticut River Valley."

A *Eubrontes* trackway parallel to the crests of ripple marks in the layer just below the main track surface shows that the dinosaur walked along the shore of a shallow lake. Try stepping on a muddy surface to make a track and you will see a real mess! The muddy sand perhaps had a thin film of algae to help preserve the prints.

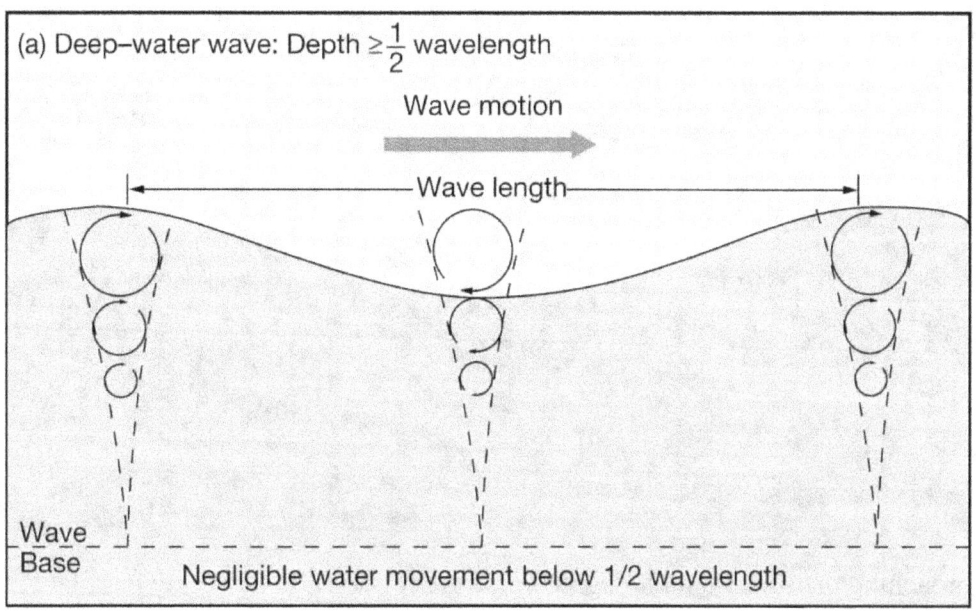

Deep-water waves transition to shallow-water waves as the waves feel the bottom at one-haft the wave length of the waves. Courtesy of the University of Indiana and Prentice Hall, Inc.

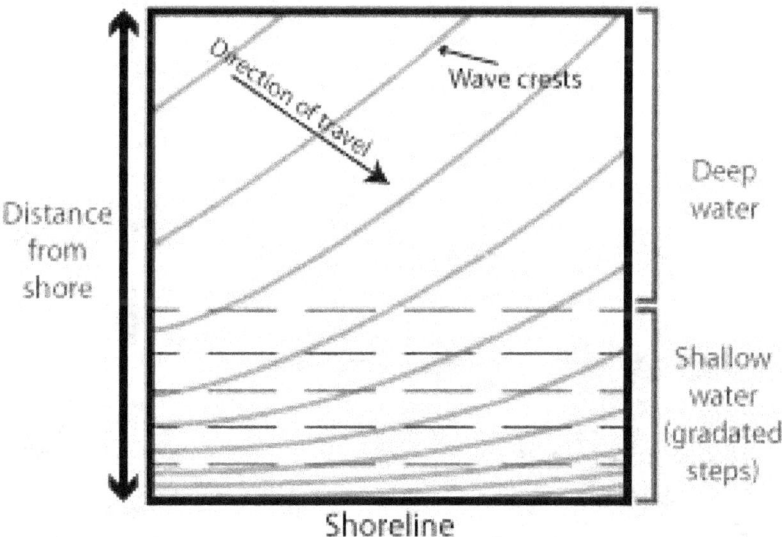

On feeling the bottom, a wave drags and slows down. The part of the wave in deep water maintains its speed and trends increasingly to parallel the shore. Courtesy of Tsunami Science.

Waves approaching the shore at a 18-degree angle are refracted to be nearly parallel to the shore. Courtesy of Credo Reference.

In 1976, the British zoologist R. McNeil Alexander used elephants, birds, people, and many other animals to formulate an equation relating an animal's speed to its hip height. His formula for calculating the speed of the animal was later modified several times and the calculated velocity lacks confidence limits, which probably are large.

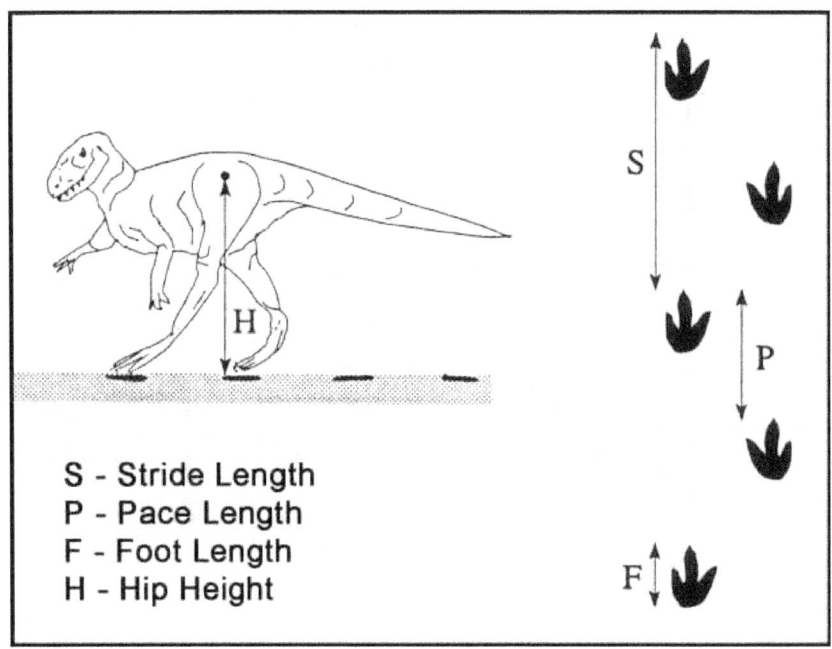

Courtesy of the City University of New York at Brooklyn.

The following simple method estimates the speed of the dinosaurs (Fallow and Galton, 2003; Lepore, 2006). Alexander found that hip height is about four times the length of the hind footprint for dinosaurs, whether bipedal or quadrupedal.

Track length, say 20 cm.

Stride, 70 cm.

Hip height, 4 x 20 = 80 cm.

Stride/hip height = 70/80 = 0.88.

If the stride/hip height is < 2, the animal was walking. If the ratio is between 2 and 2.99, trotting, and over 3, running. This hypothetical maker of a *Eubrontes* trackway was ambling along.

At Route 5, the dinosaurs were walking about 3 to 5 miles per hour, comparable to walking speeds of humans (Farlow and Galton, 2003; Lepore, 2006). The critters were not in a hurry. Study of many sites with dinosaur trackways demonstrates that preserved dinosaur speeds are low, between 2.2 to 8 miles per hour.

After viewing the track site, you can walk down to the Connecticut River to see a lacustrine deltaic sandstone interbedded with deeper-water gray mudstone. The sandstone is stratigraphically above the dinosaur tracks. The currents that built the delta flowed to the southwest, roughly parallel to the trend of the Connecticut River.

Fragments of plants in the sandstone suggest what the herbivorous dinosaurs were eating. The web site of the Lyman Plant House and Conservatory notes that the Triassic-Jurassic plants were dominated by gymnosperms, which are seed-producing plants such as conifers, cycads, and ginkgos. Although there are many living conifer and cycad species, the ginkgo is the only species of a primitive gymnosperm group that flourished in the Jurassic. Lyman Planthouse has specimens of primitive cycads that would be as tough to eat as cardboard. The planthouse is at 16 College Lane on the Smith College campus in Northampton, Massachusetts.

Today the plant world is dominated by flowering plants (angiosperms), which first flourished about 140 million years ago in the Lower Cretaceous.

3. Playa and Lacustrine and Strata, Holyoke Community College

What to see. - Playa redbeds are interbedded with two sequences of lacustrine gray strata in the lower Jurassic East Berlin Formation. Characteristic playa features include ripple marks, mudcracks, layers of muddy dolostones, and disruption of bedding fabric by repeated wetting and drying. The lacustrine strata include ripple-cross laminated sandstone deposited by sheet floods.

How to get there. - From I-91 take exit 16 and proceed west on Route 202 (Homestead Road) for one mile to the Holyoke Community College. The outcrop is just inside the entrance to the HCC campus opposite parking lot A.

Comments. - Shallow perennial lakes were present during the wetter part of the Earth's precession cycle. During the dry phase of the cycle, playa redbeds replaced the lakes because precipitation was inadequate to sustain a lake.

The paleocurrent directions of playa and lacustrine sandstones in the East Berlin Formation show that sediment transport was consistently eastward from the hinged (unfaulted) western margin down the tilted floor of the rift valley to the toes of alluvial fans along the eastern border fault. The fans were small with radii of a few kilometers. The sheet flows were moving down the degree-or-so slope of the playa surface toward an elongate north-south, topographically-low area created by subsidence along the border fault (McDonald and LeTourneau, 1990; Hubert et al., 1992). The shallow lakes deepened eastward toward the fans and the lacustrine strata thicken towards the fault. As precipitation varied, the lakes expanded and contracted, surrounded by extensive mudflats where dinosaurs and other reptiles left their prints and trackways

Overlying playa redbeds, the bottom of the upper (85-cm-thick) lake sequence is at the top of the meter stick. The top of the lower lacustrine sequence (40-cm-thick) is behind my hat. The white sandstones near the top and bottom of the lake sequences are sheet-flood sandstones. The lacustrine strata are gray because the original yellow-brown limonite soil stains on the surfaces of the grains were removed by organic acids in the lake. Limonite is preserved in the playa sediments and over hundreds of thousands of years dehydrates to hematite. The thin layers of brownish-red playa deposits within the lake sequences demonstrate the ephemeral nature of the shallow lakes. Photo by Jim Dutcher.

Similar to the East Berlin Formation, alluvial fans lead down to a flooded playa (Owen's Lake) in California. The white is an ephemeral salt crust. Courtesy of Maven's Photoblog..

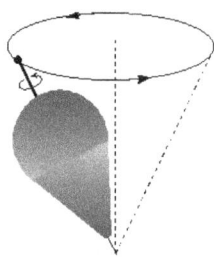

Like the Earth, the axis of a spinning top traces the surface of a cone. Courtesy of NASA.gov.

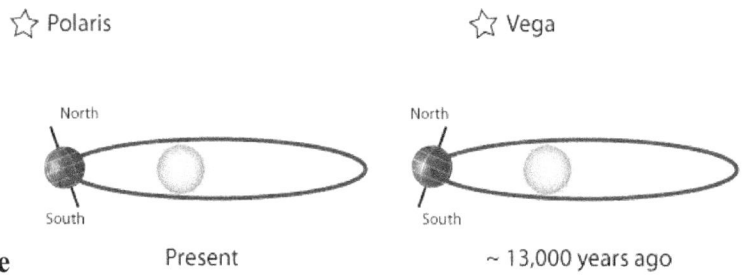

The pull of the Moon and the Sun cause the Earth's axis to wobble, producing the precession cycle of 26,000 years. Today the Earth's axis points away from the sun towards the Polaris star. Half a cycle ago, the axis pointed towards the Star Vega in the constellation Lyra. Lakes were present in the northern hemisphere when it was closer to the sun because for each one-degree F increase in temperature the atmosphere can hold four percent more water, increasing precipitation. Today the precession cycle is 26,000 years, but over time the cycle slowly lengthens so that back in the early Jurassic it was 20,000 years. From the middle of the lower lake sequence to the middle of the upper is 192 cm. A rough estimate of the sedimentation rate is 10 cm per 1,000 years. Courtesy of Reasons to Believe.org.

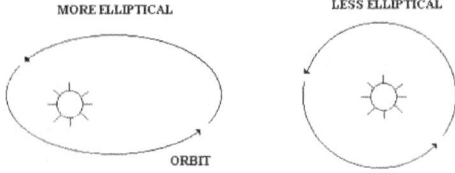

The thickness and clustering of the lake beds in the East Berlin Formation also show the effects of the orbital shape and axial tilt of the Earth (Olsen *et al.*, 2011). The axial tilt is currently 23.4 degrees and varies from 22.1 to 24.5 degrees in a 41,000-year cycle. Due to the pull of Jupiter and Saturn, the orbital shape varies from nearly circular to mildly elliptical in a cycle of 400,000 years. Courtesy of Indiana University.

The one-centimeter-thick layers of muddy dolostone reflect precipitation of Mg-rich calcite (calcium carbonate) by intense subtropical evaporation of temporary lakes. After burial the calcite was converted to dolomite (Ca/Mg carbonate). Iron substitutes for some of the Mg in dolomite so it weathers to yellow-brown limonite (hydrous iron oxide). The strata above the muddy dolostones contain dolomitic layers with a disturbed fabric.

The light-colored layer is ripple cross-laminated, fine-grained sandstone (4-cm thick, middle of the photo). A flood event filled a shallow swale (left side) eroded into the underlying lacustrine gray mudstones. Above the sandstone is a yellow-weathering, muddy dolostone that desiccation broke up into angular pieces. Photos by Jim Dutcher.

Three layers of playa redbeds have disrupted fabrics due to repeated flooding, with cycles of water-saturated mud followed by evaporation and drying. Precipitation and dissolution of salt helped churn the sediment. In the middle layer, lumps of dolomitic mudstone are dispersed in brownish-red mudstone.

A sediment-laden flow decelerated on the playa surface, depositing cross-laminated ripples with flow to the left. The individual ripples are one-centimeter thick. The two beds below the ripples are thoroughly disrupted by repeated wetting and drying. The lowest bed retains a few remnant patches of ripple cross-lamination. Photos by Jim Dutcher.

Looking up through the beds, you can see the characteristic playa association of ripple-marks and mudcracks. Photo by Jim Dutcher.

In southwest Idaho, clumps of peppergrass thrive when the playa is seasonally flooded. Mudcracks are commonly present with dinosaur tracks in playa redbeds. Courtesy of Thayne Tuason and Wikipedia

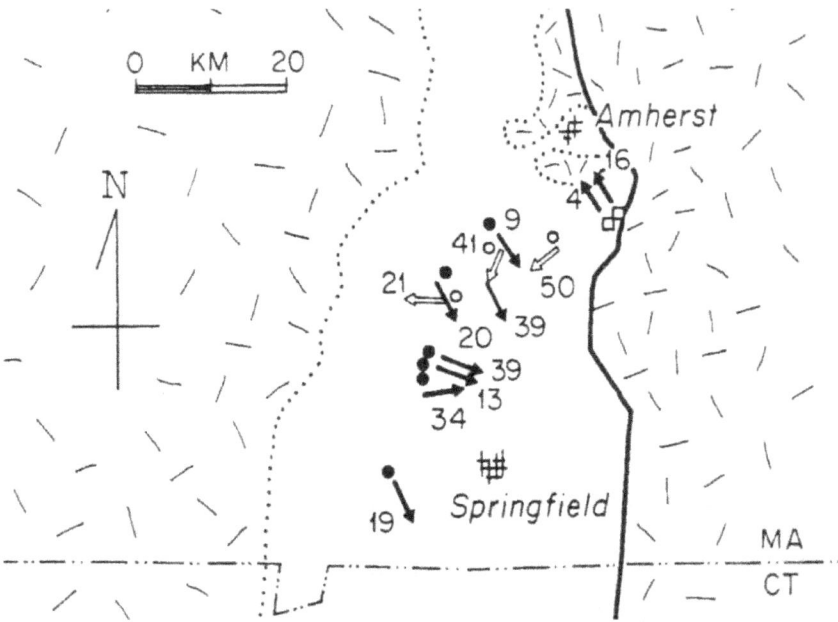

Paleocurrent directions for redbeds in the East Berlin Formation in the northern Hartford basin. Solid circles are playa sandstone and siltstone; open circles are fluvial sandstone; open boxes are alluvial-fan pebbly sandstone. Holyoke Community College is located close to and just east of the three vertical locations located northwest of Springfield. Floods from highlands west of the basin flowed downslope to deposit muds and sands on the playa and margins of the lakes.

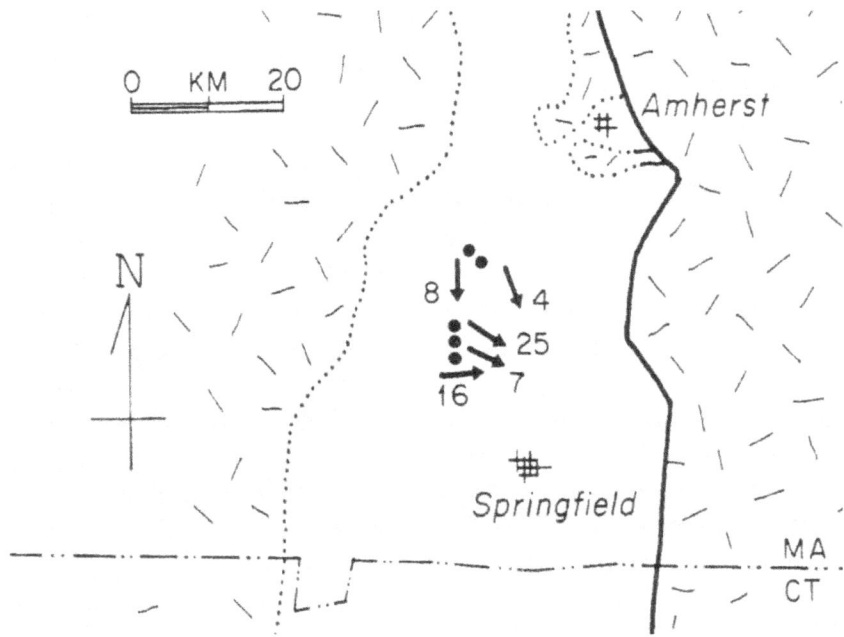

For the lacustrine strata, the black dots are sandstones and siltstones From Hubert *et al.*, 1992.

4. Dinosaur Park and Arboretum, Rocky Hill

What to see. - The 55,000-square-foot geodesic dome houses more than 500 tracks and trails, mostly *Eubrontes*, in the East Berlin Formation plus dioramas of Triassic and Jurassic life and environments, interactive exhibits, a lecture hall, and a gift shop. The park comprises 80 acres with two miles of nature trails featuring 250 species and cultivars of conifers, as well as katsuras, ginkgoes, magnolias, and other surviving representatives of plant families from the Triassic and Jurassic.

The tracks were discovered in 1966, when a bulldozer operator was excavating the foundation for a state office building. The building was built at another site and the Dinosaur Park and Arboretum opened in 1968, the same year that it was declared a National Landmark by the U.S. Department of the Interior. It is also a World Heritage site. The park is managed under the Bureau of Outdoor Recreation, Connecticut Department of Energy and Environmental Protection.

In addition to the tracks under the dome, another 1,500 tracks are buried nearby for preservation.

How to get there. - The park is one mile east of Exit 23 on I-91, a few miles south of Hartford, Connecticut. The address is 400 West Street, Rocky Hill, Connecticut. The Exhibit Center is open Tuesday-Sunday 9:00-4:30 a. m. The hiking trails are open Tuesday-Saturday 9:00 a. m.to 4:30 p. m. The trails close at 4:00 p.m. The daily admission fee for adults (ages 13 and up) is $6, youth (ages 6-12) is $2, and children under 6 are free. Connecticut seniors over 65 are eligible for a Charter Oak Pass.

Comments. - During the early Jurassic 200 million years ago, the muddy sandstones at the park accumulated on a playa located on the floor of the rift valley. A playa is a desert basin with no outlet that during storms partially fills with water to form a temporary lake.

Visitors examine the tracks on the original discovery surface. The geodesic dome houses the Exhibition Center. Courtesy of the State of Connecticut and Connecticut Museum Quest.

The *Dilophosaurus* diorama and the side-lighted tracks are inside the dome. The tracks are mostly *Eubrontes*, but also present are tracks of the dinosaurs *Grallator* and *Anchisauripus*, and the crocodylomorph *Batrachopus* (Lull, 1953). Courtesy of Yelp.

The type specimen of *Eubrontes* is at the Beneski Natural History Museum of Amherst College. In 1991, *Eubrontes* was named the Connecticut state fossil. The scale is 5 cm. Courtesy of Paul Olsen. *Dilophosaurus* is inferred to be the *Eubrontes* track maker. At 23 feet long and weighing 880 pounds, *Dilophosaurus* was one of the largest carnivorous dinosaurs of the lower Jurassic. Courtesy of Dinosaurpictures.org.

Directional paleocurrent readings of ripple marks and cross-beds at and near the park show that streams from highlands on the west side of the valley spread layers of muddy sand over the playa. During the rainy season, a shallow lake was present in the lowest part of the playa. The alluvial fans on east side were relatively small and not important sediment sources.

The many trackways and individual prints were made over a substantial period of time because they have no preferred overall orientation (Galton and Farlow, 2003). Also, the tracks vary in depth and sharpness. The animals were visiting a shallow lake to get water and catch fish.

The surface with the tracks has a number of marks interpreted as made by swimming dinosaurs (Coombs, 1980). An example is adjacent to the wooden walkway where a large magnifying glass allows visitors to closely examine the marks while standing on the walkway. The inference of swimming is controversial because some observers interpret the marks as unusual walking prints. You can decide looking at the evidence.

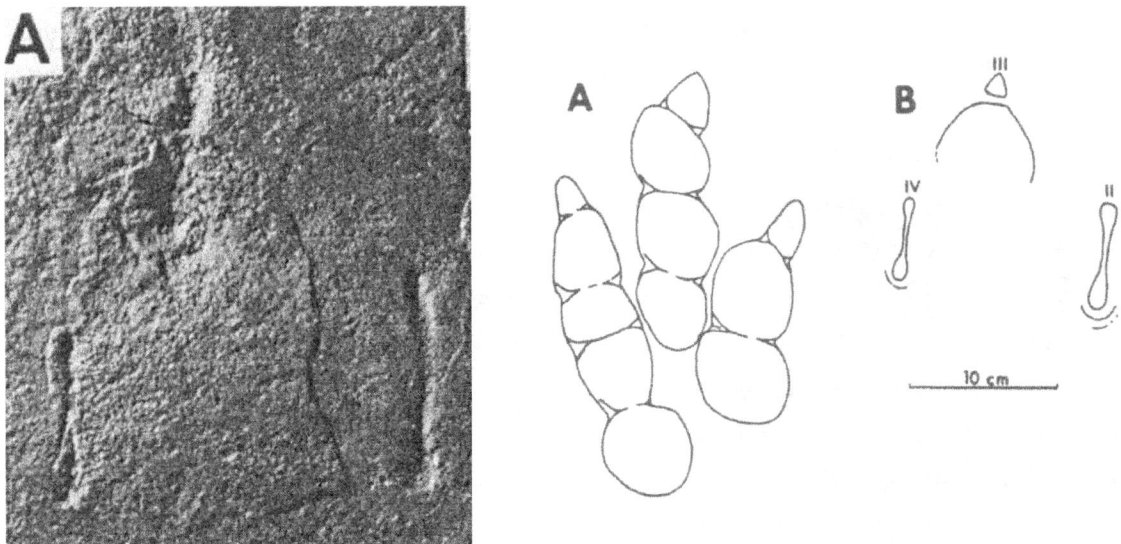

Possible *Eubrontes* swimming print with clear traces of the lateral claw scratches. On the right is a *Eubrontes* print and a composite sketch of several swimming prints. Courtesy of Walter Coombs.

Interpretation of a swimming theropod dinosaur making scratch marks. The dinosaur is 9 m long from nose to end of tail. Sketch by Matthew Hyman. Courtsey of Walter Coombs (1980).

In 1990, the Friends of Dinosaur Park and Arboretum commissioned William Sillin to paint "*In the Late Triassic*" for the Exhibit Center. You are looking south with fault-bounded highlands on the east side of the rift valley. The tall conifer *Araucarioxylon* grows in the sandy soil at the toes of the alluvial fans. The 7-m-long phytosaur *Rutiodon* is in the lower left. The 6-m-long dinosaur *Dilophosaurus* is in the lower center. Courtesy of William Sillin and the Dinosaur Park and Arboretum.

The *Eubrontes* track maker is thought to be the carnivorous dinosaur *Dilophosaurus* because bones of *Dilophosaurus* are associated with *Eubrontes* tracks in lower Jurassic strata in Arizona. At Rocky Hill, the tracks range from 10 to 16 inches long and the prints in trackways are spaced 3.5 to 4.5 feet apart.

Dilophosaurus was a meat eating, bipedal dinosaur with two rounded thin crests on its skull, which gives the dinosaur its name (two-crested-reptile). The crests are made of extensions of bones and were initially thought to be for attracting mates or intimidating rivals, but the current view is they were for intra-species recognition.

Although *Dilophosaurus* was among the largest carnivores of its day, it scavenged carrion because the sharp, pointed teeth were too weak to bring down large prey. The shape of the teeth would allow it to catch fish, which matches the common occurrence of *Eubrontes* tracks in playa and lacustrine strata.

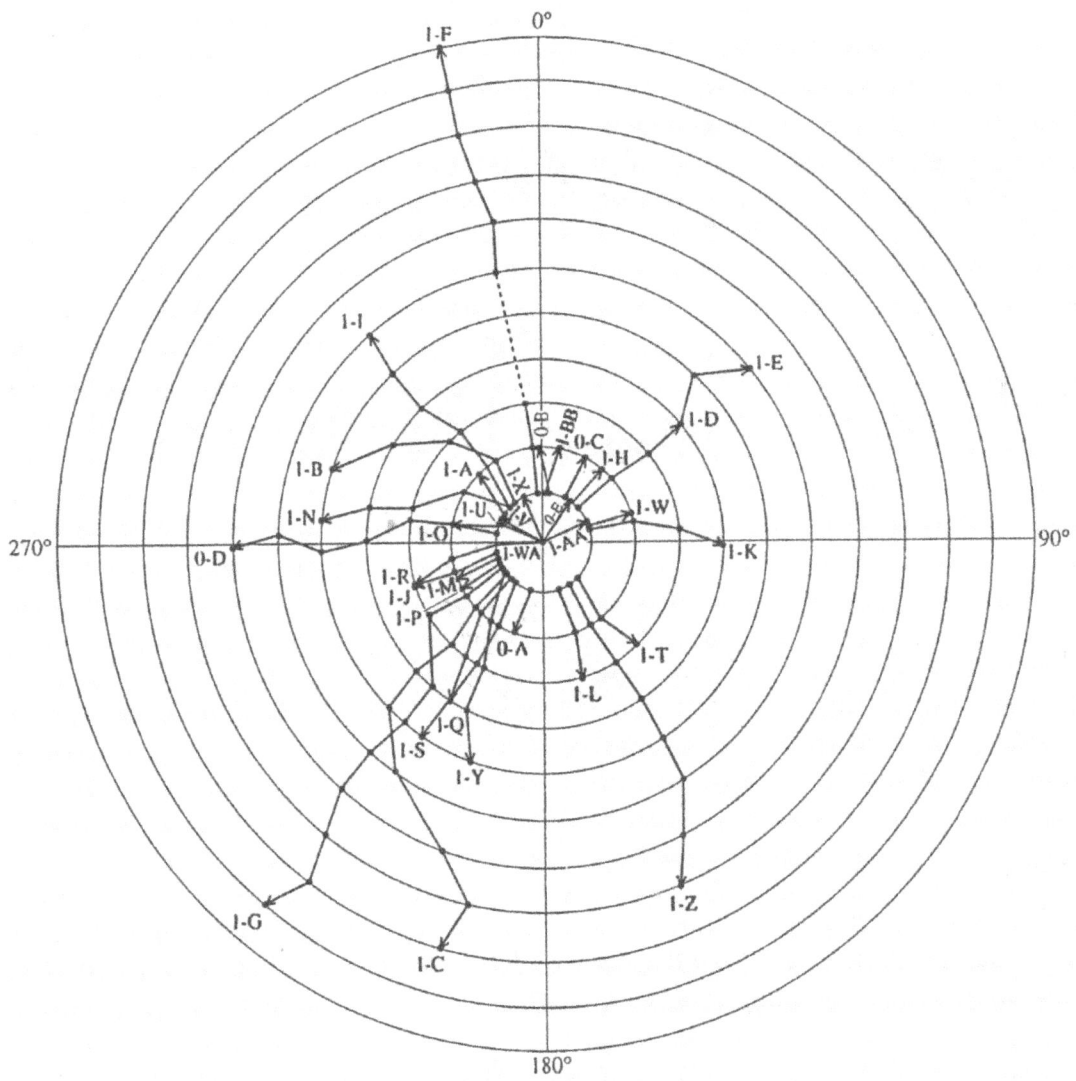

Direction of travel of the 28 dinosaur trackways plotted with the start of each trackway in the center of the circles. Magnetic north is zero. The trackways do not show any preferred direction of movement. Some of the dinosaurs evidently rambled as they walked along. Courtesy of Farlow and Galton and Columbia University Press.

Farlow and Galton (2003) write, "There is no obvious indication that the Dinosaur State Park tracks were made by a large group of animals moving through the area together. Our impression is rather one of individual dinosaurs or perhaps of small groups of dinosaurs passing across the site at different times. Osborn (1972) came to the same conclusion. The footprint-bearing surfaces may have recorded tracks over a long period of time – perhaps weeks or months."

Dilophosaurus strolls in a park. Courtesy of Mark Mancini and mentalfloss.com.

Field Trip in the Amherst Block

5. Beneski Natural History Museum, Amherst College

What to see. - Opening in 2006, the Beneski Earth Sciences Building houses the natural history museum and the Department of Geology. The famous collection of dinosaur footprints and trackways is housed on the ground floor. The rock slabs on display also include other reptile, amphibian, and invertebrate impressions, ripple marks and raindrop splash marks.

Large ice-age mammals, including mastodon, mammoth, and Irish elk are displayed on the first floor, which also has an exhibit on the evolution of the horse in North America. The second floor holds invertebrate fossils, trace fossils, minerals, and exhibits on local geology.

Beginning in 1835, Professor Edward Hitchcock of Amherst College amassed a collection of more than 21,000 specimens with footprints, including many gathered by Dr. James Dean, Roswell Field, Dexter Marsh, and workers Hitchcock hired. Edward Hitchcock joined the faculty of Amherst College in 1825 and was the third President from 1845 to 1854.

More than 1,700 fossils, rocks, and minerals are on display in the three floors of exhibits. The specimens may not be touched. Tens of thousands of specimens are available for use by scholars and researchers from around the world. The collection dates back to the earliest days of the college with some 200,000 objects.

The museum is in the ground, first, and second floors. The third floor is restricted to students and staff of the Department of Geology.

"Amherst 190 million years ago" by William Sillin is displayed in the museum. Photos courtesy of Amherst College and William Sillin.

Clockwise from upper left. 1) The dinosaur *Allosaurus*. 2) Side-lighted dinosaur tracks. 3) Large ice-age mammals and other beasts. 4) Closer view of the ice-age mammals. A wooly mammoth is on the right. The mammoth is now the mascot of Amherst College. The strange antlers on the upper left are those of an Irish elk.

A few of the minerals on display. Photos courtesy of Amherst College.

Clockwise from upper left. 1) Tracks of the dinosaurs *Anchisauripus* and *Grallator*. 2) Cross-section of extinct marine ammonite. 3) Students listen to a talk. 4) Amherst College expedition of 1911 to the American west with fossils wrapped in plaster. Courtesy of Amherst College.

How to get there. - The museum web site gives directions to travel to the museum. "Take Interstate 91 north to Exit 19 in Northampton; take Route 9 east across the Calvin Coolidge Memorial Bridge to the center of Amherst. Continue on Route 9 east 3/10ths of a mile. Just before the purple-and-white railroad bridge, take a right onto (unmarked) East Drive and walk past the campus police building. At the stop sign, turn right up (unmarked) Barrett Hill Road. The museum is a red brick building with a metal roof. (Note: The College does not have public parking on weekdays. Visitors can park in the lots or in the garage in the town center and can easily walk to the museum.)"

A wheelchair and walker are available for use at no cost while visiting the museum. The museum is open Tuesday to Friday, 11 a.m. to 4 p.m., and Saturday and Sunday, 10 a.m. to 5 p.m. It is closed Monday.

Noah's Raven uncovered 1801 or 1802, in South Hadley, Massachusetts. Courtesy of dinotracksdiscovery.org.

Comments. - An historically important specimen is "Noah's Raven," the first evidence of a dinosaur in North America. The trackway is from the Shuttle Meadow Formation. The story of Noah's Raven is told at the web site of the Nash Dinosaur Track Site and Rock Shop located in South Hadley, Massachusetts.

"The first dinosaur tracks in recorded history were discovered by a young farm boy by the name of Pliny Moody. At about age 12, he was plowing a field near Moody Corner, South Hadley, Massachusetts, when he discovered a slab of rock with stone tracks on it. It is said that he then took the slab of tracks home and installed it as a door step at the family home. A few years later, about the time Pliny Moody went off to school, the tracks were bought by Dr. Elihu Dwight of South Hadley, Massachusetts. Dr. Elihu Dwight was the first person in recorded history to purchase a dinosaur track. While in the possession of these two men, they obtained the name "the tracks of Noah's raven" because the biblical Noah, when he was on the ark, released a raven that never returned. It was believed that the raven landed in South Hadley, Massachusetts, and left its tracks in the mud, which later turned to stone. After obtaining the tracks, Dr. Dwight had the tracks in his possession for about 30 years. About 1839 they were obtained by Professor Edward Hitchcock for Amherst College. The slab of tracks is known as 16/2 in the Amherst College collection."

Field Trips in the Deerfield Basin

6. Summit of Sugarloaf Mountain, South Deerfield

What to see. - The summit at 791 feet provides an outstanding vista of the Connecticut River, Pioneer Valley, and Pelham and Berkshire hills. A pavilion for picnicking is at the summit.

How to get there. - Going south on I-91, take exit 25, or going north take exit 24, and proceed east on Route 116 for about a mile. The state park is next to the stop light on Route 116, just west of the bridge over the Connecticut River that connects South Deerfield to Sunderland. The entrance to the park is at 300 Sugarloaf Street. The park is open daily from 9 a. m. to sunset. The daily parking fee is $5 for a MA vehicle and $6 for out of state visitors. A narrow auto road that winds to the summit is open starting May 1. Parking is available at the summit.

Comments. - The Sugarloaf Arkose is an alluvial-fan deposit, part of a bajada built by rivers that flowed from highlands located west of the rift valley. The bajada formed when alluvial fans coalesced on the hanging wall (opposite the faulted side) of the half graben. The formation of a bajada is favored by 1) a monsoonal climate, 2) lack of vegetation, 3) floods with high sediment yields, and 4) wide fluctuations in discharge peakedness (Leier *et al.*, 2005). In the Deerfield rift basin, the dry-dominated monsoon suggests these four factors were present.

The outcrops of the Sugarloaf Arkose indicate a bajada of at least 625 km^2. The size of the Sugarloaf bajada was limited by the size of the subsiding basin, which was substantially larger than a typical semi-arid alluvial fan, but smaller than many modern bajadas.

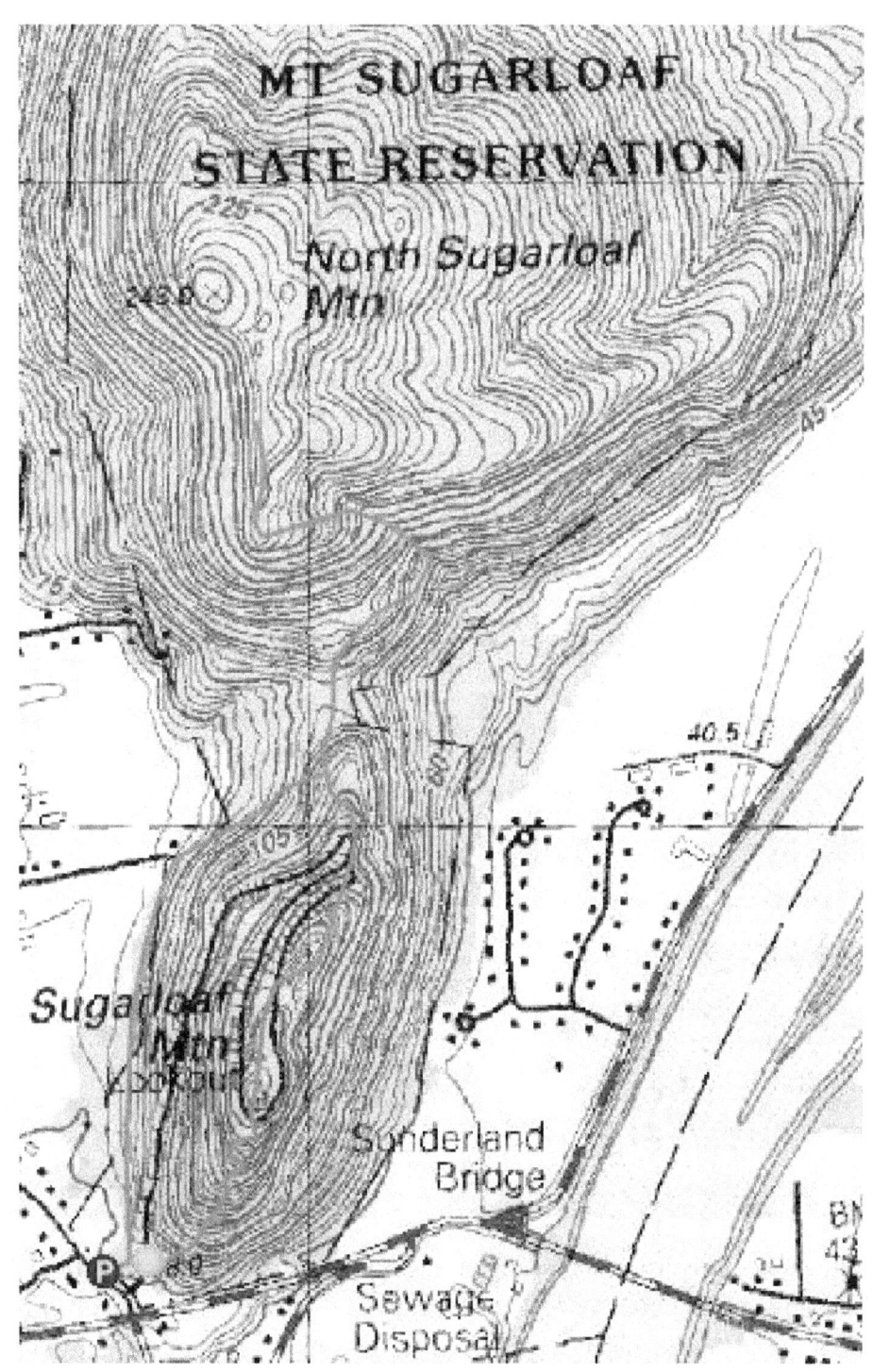

Map of part of Mount Sugarloaf State Reservation. The summit overlook at 652 feet is on the south end of South Sugarloaf Mountain. The east-west highway south of South Sugarloaf is Route 116. Courtesy of FranklinSite.com and the U.S. Geological Survey.

Marked by blue blazes, the Pocumtuck Ridge trail extends 15 miles from Sugarloaf Mountain in South Deerfield (lower left) northward to Rocky Mountain with Poet's Seat Tower in Greenfield. The summit is made of sandstones whereas the Deerfield Basalt is exposed north of the summit. The low terrane south of the Sugarloaf Mountain is relatively soft, reflecting the precipitation of illite clay-mineral cement and illite replacement of the former albite cement during a "hot spot" hydrothermal event. Courtesy of TerraMetrics and the USDA Farm Service Agency.

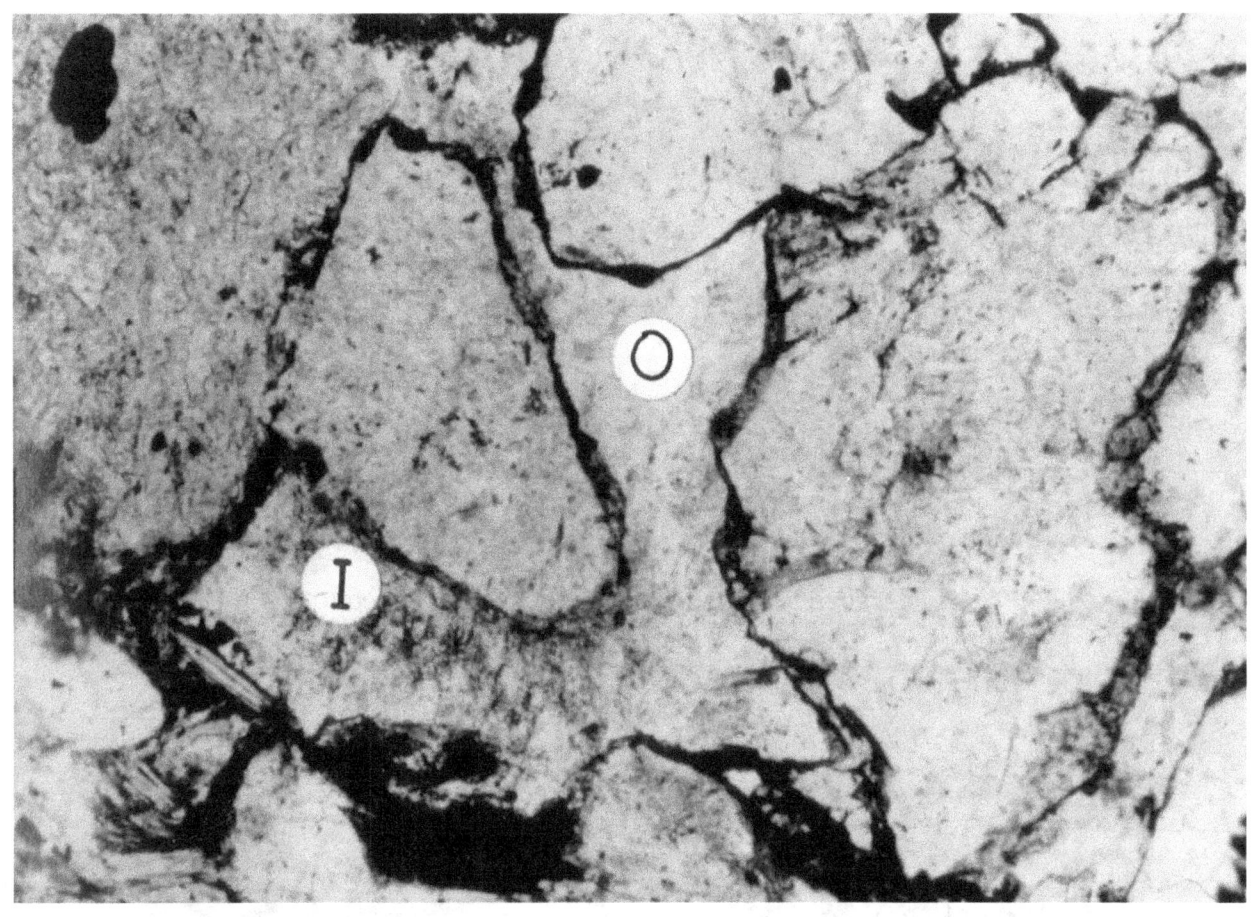

Sugarloaf Mountain is unusual because the summit is made of sandstone and not the nearby Deerfield Basalt, which one would expect at the summit. The sandstones are more resistant to weathering and erosion than the basalt because they are cemented by albite that is almost as hard as quartz and is chemically more stable than Fe-bearing minerals in basalt such as hornblende, augite, and olivine. Albite is a sodium-rich aluminum silicate mineral. This thin-section photomicrograph of Sugarloaf sandstone shows a typical albite overgrowth (O) on the plagioclase sand grain adjacent and to the right of the overgrowth. Albite overgrowths are also present on the right-hand side of the plagioclase grain. Unidentified small inclusions (I) are present in the albite cement. The heavy dark lines around the sand grains are the iron-oxide mineral hematite that gives the Sugarloaf sandstones their characteristic brownish-red color. The long edge of the photo is 1.12 mm. Courtesy of John Taylor.

Picnic area near the tower. Courtesy of Wikipedia.

Smaller and larger trough cross-bed sets are adjacent and just below the scale, which is divided into centimeter and inch divisions. You are looking down the axes of the concave-up trough sets with river flow into the rock face to the southeast. The outcrop is between the picnic area shown above and the tower. The formation of trough cross-beds sets is included in the comments on the outcrop on Country Club Road.

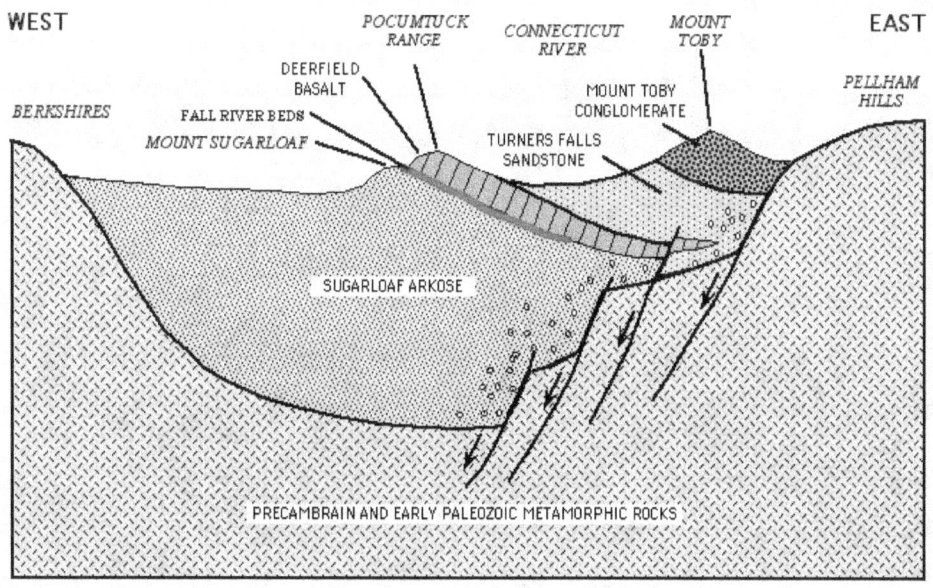

Cross section of the Deerfield basin just north of Amherst. Courtesy of Planetward.org

Looking east from the summit, the sloping flat surface of Long Plain Brook delta is visible at the top center below Mount Toby (the highest peak) and above the bridge. Long Plain Brook constructed the delta on entering glacial Lake Hitchcock, which existed from 15,500 to 12,500 years ago. The field trip to see the Mount Toby Conglomerate at Roaring Brook is on the far side of Mount Toby. Courtesy of elevation.maplogs.com.

Looking south from the tower overlook, you see the jagged outline of the Holyoke Range comprised of the Jurassic lava flows of the Holyoke basalt. The Connecticut River Greenway Park of Massachusetts is gradually acquiring the river banks, including the levees visible here. The post-glacial, meandering Connecticut River is depositing channel sands and floodplain muds on a surface cut into the deposits of glacial Lake Hitchcock, so that the river deposits cover the lake deposits.

In the middle distance (upper right) is Mount Warner, a resistant knob of Lower Paleozoic metamorphic rocks. Mount Warner and the Holyoke Range were islands in the glacial lake. The flat upper surface of the ice-contact Hadley delta adjoins Mount Warner on the left side with the ice-contact steep slope of the delta facing to the left (northeast). The delta was built by water flowing into the lake from a tunnel in the ice. The stable level of the glacial lake was half way up the red brick University Library just visible in the far upper left corner next to the white Lederle Graduate Research Center. You can visualize the lake level by connecting the delta surface with an elevation half-way up the university library. Or you can extend the Long Plain Brook delta surface to the Mount Warner delta.

Buried post-glacial, river-channel sands are locally present beneath the floodplain muds. Just east of Mount Warner, a buried river channel of the Connecticut River is an important source of groundwater for the town of Hadley. The barns in the fields to the left of the river are for curing shade-grown, broad-leaf tobacco leaves used to wrap cigars. Courtesy of John Burk.

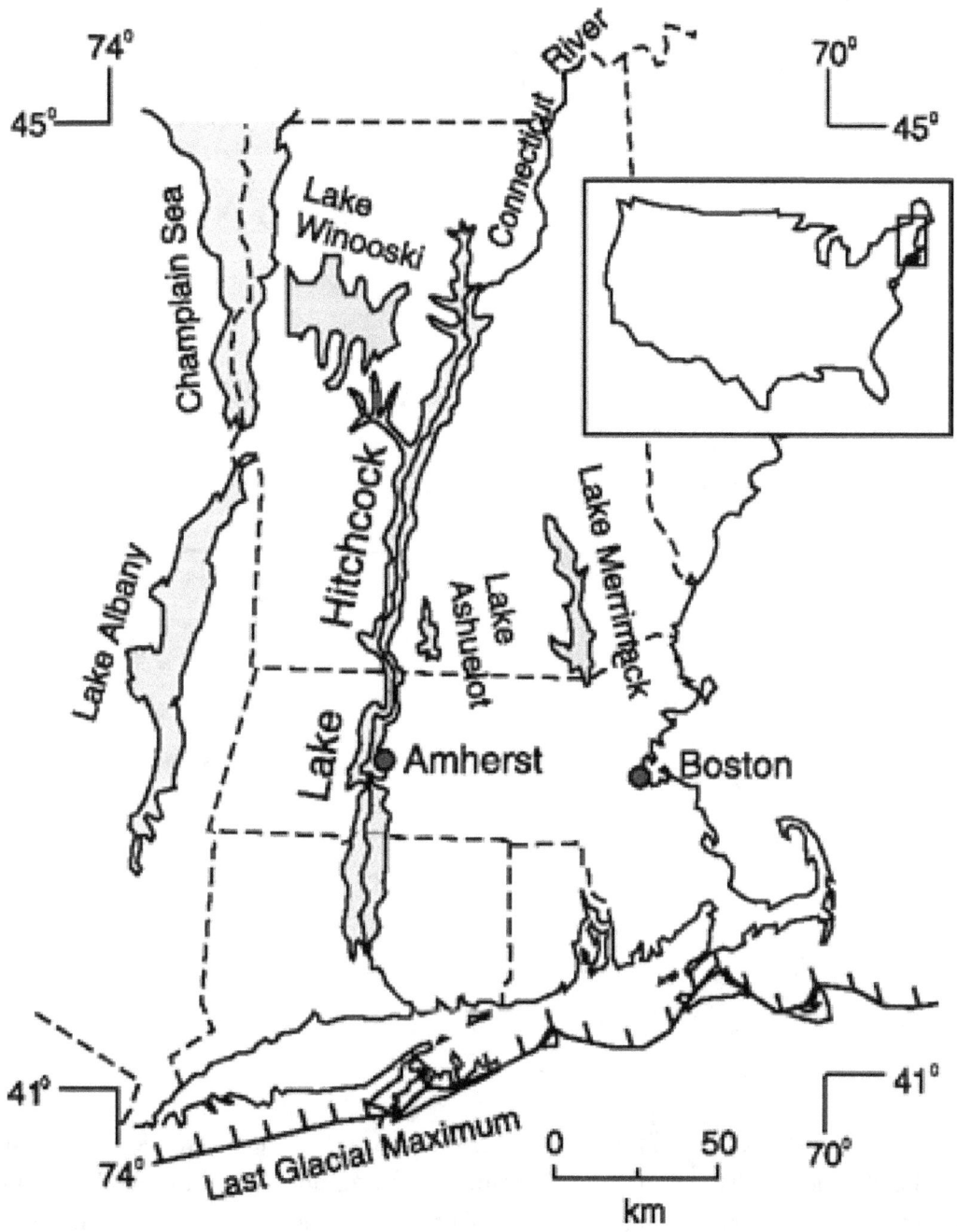

. At its maximum extent, glacial Lake Hitchcock extended 320 miles from a glacial moraine dam at Rocky Hill, Connecticut, to Saint Johnsbury, Vermont. Lake Hitchcock drained when the outlet stream at Rocky Hill eroded through the moraine dam. Courtesy of Instaar.Colorado.edu.

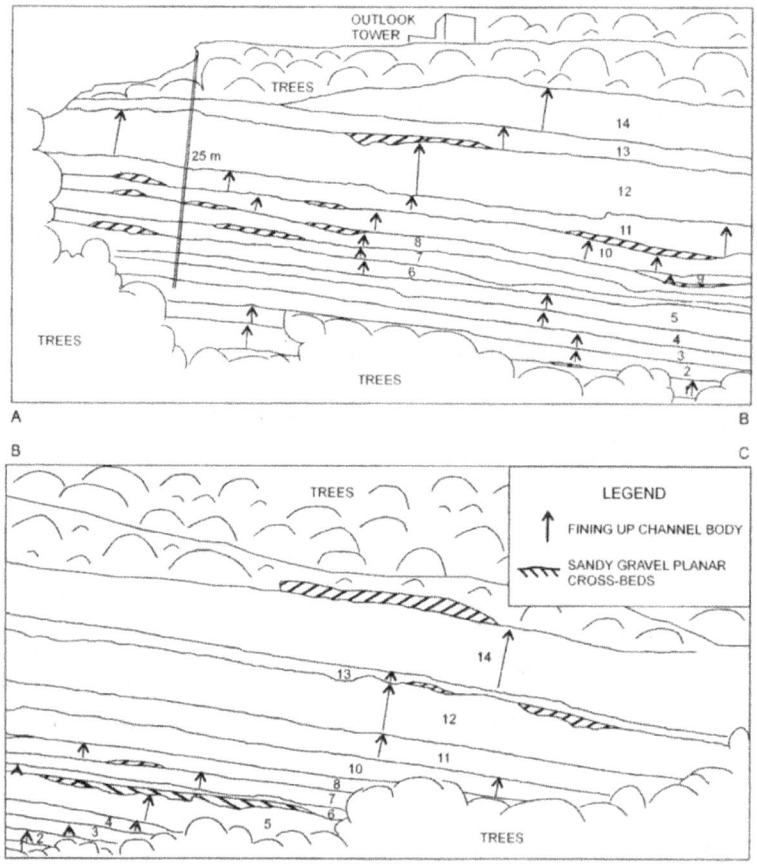

The observation tower is visible at the top of the cliff (left center). The sketch is of the cliff face. The channel sandstones are mostly grain-size fining-up with minor overbank mudstones. The biggest boulders exceed 40 cm in size; channel depths are 2.7 to 4.5 m. The entire cliff is a slice through an alluvial fan built by a river that flowed from the west side of the subsiding basin. Average paleoflow in the 14 channel sandstones is out of the cliff face over the viewer's right shoulder towards 88 degrees, at an angle of 57 degrees from the cliff face. The paleocurrents were measured along the road that leads to the tower. From Hubert *et al.*, 2008.

The cross-beds used to measure the direction of river flow are exposed along the road to the summit. The upper photo is a longitudinal section through two subaqueous dunes with flow to the left. In the lower photo, A) points to transverse cross sections of trough-shaped cross-bed sets; B) is a cross section through a sand bar that build out from the river bank located at C. Weathering has deteriorated the outcrops since the photos were taken in 2005.

On the road to the summit, I hold a meter stick next to a cross-bedded subaqueous dune of pebbly sand with flow to the left.

Conglomerate and pebbly sandstone are exposed at the rear of a flower garden adjacent and south of the tower.

Present in a closer view of the conglomerate are vein quartz, granite, and the metamorphic rock schist. With the same composition, larger cobbles round faster than smaller pebbles over the same distance of transport.

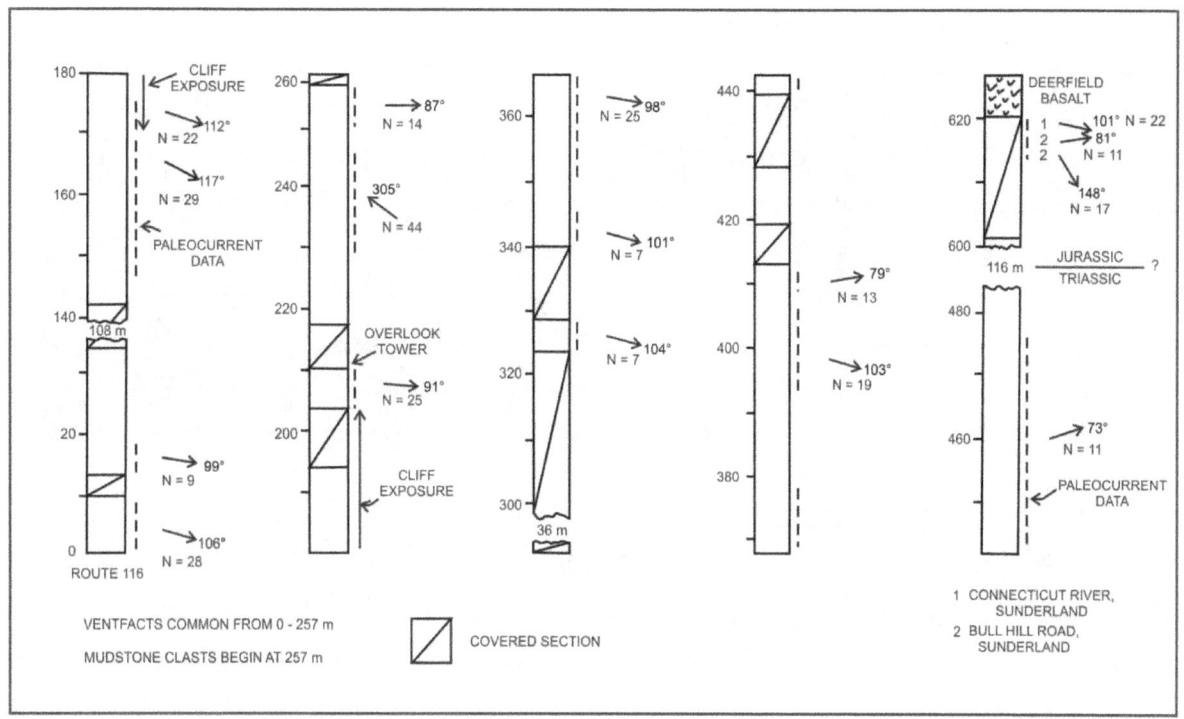

Measured section and paleocurrents of the Sugarloaf Arkose starting at the exposure on Route 116, proceeding up the road to the summit, and then on to the Deerfield Basalt. River flow was mostly towards the southeast and east. The vertical scale is in meters. From Hubert *et al*., 2008.

From 0-257 m in the measured section, the sequence consists of river-channel pebbly sandstones interbedded with minor overbank mudstones. This is the upper part of the fan where there are active and abandoned channels. During floods, mud by-passes the area. The semi-arid climate supported relatively few plants to stabilize the channel banks, which favors channel avulsions where a new channel is created by a rapid lateral shift of the channel to a new path down the fan, commonly eroding down into the underlying channel sands.

Above 257 m, thin beds of red mudstone become more common and the conglomerates contain a few percent of mudstone clasts. As the basin floor subsided, the fan advanced eastward while expanding westward. The highlands to the west were never far away as shown by the large size of the crystalline boulders, some exceeding half a meter.

Gravelly planar cross-beds thicker than 40 cm are gravel bars deposited in scours in river channels. During high-river stages, the flow entered a scour, slowed, and constructed a bar of pebbly sand that advanced into the scour.

At the summit, Tom Ricardi explains how he raises bald eagles from eggs while holding an eagle brought back to health after an injury. Tom is a wildlife rehabilitator who in 1975 founded the Massachusetts Birds of Prey Rehabilitation Center. On the right, the box with the young eagle is covered while the crowd forms two lines that lead to the cliff edge.

The young eagle on its first flight, seconds before a Peregrine falcon buzzed it, suggesting the eagle find a home somewhere else. Photos by Jim Dutcher.

On inactive parts of the fan between channels, many stones were sculptured by wind-driven sand into ventifacts, some with dark desert varnish composed mostly of fine-grained clay minerals. The clays are mixed with black manganese oxide and red iron-oxide. During flood events, the channel banks collapse and the ventifacts are swept down channel. The proportion of ventifacts in channel gravels varies from scarce to abundant, reflecting local availability.

The summit is commonly used as a release point for birds raised from eggs or that have undergone rehabilitation. On August 17, 2005, at about 10:00 a. m., Jim Dutcher and I were studying the outcrops along the road up the mountain, debating whether a cylindrical structure four inches in diameter and at least a foot long was a concretion or a burrow of an early mammal or dinosaur that lived along the bank of a Late Triassic river. A steady stream of cars and small trucks began passing us on the way to the parking lot at the summit, including a van labelled *Hampshire Gazette*, our local newspaper. We trudged the 100 or so vertical feet to the top and joined the crowd.

A pickup truck arrived and a large wooden box was set on the ground. Tom Ricardi, the host in charge, arranged the crowd in two lines, defining a corridor that ended at the vertical cliff edge. The box was placed at the head of the corridor and Tom explained that for years he has hatched bald eagle eggs and raises the chicks to young adults ready to be set free. This was in August because eagles fledge about the end of June.

The box was on its side. The lid was removed, the young eagle hopped forward, took a few tentative steps, soared into air, and flew down the corridor and up into the sky above the crowd. Seconds later a much smaller Peregrine falcon that had a nest on the vertical rock face, dived and buzzed the eagle at terrific speed, but probably not its maximum 240 miles per hour. Like a P-51 mustang fighter in World War II, the eagle dipped its left wing and dove 530 feet down to the surface of the river and flew upstream.

In 1989, an eagle release program was started in Massachusetts. In 1900, there were no known nesting bald eagle pairs, whereas in the 2012 breeding season, 38 territorial pairs were counted. Caretakers use eagle "puppets" to feed the chicks so that they imprint on eagles rather than humans. The eagles reach breeding age at 4 to 5 years and live about 35 to 40 years in the wild. In 2007, the status of the bald eagle in Massachusetts was upgraded from endangered to threatened.

7. Fluvial Redbeds, Route 116 at Sugarloaf Mountain, South Deerfield

What to see. - The road cut on Route 116 at base of Sugarloaf Mountain exposes fluvial redbeds in the Sugarloaf Arkose deposited by a river that flowed from the west side of the basin during the late Triassic. The outcrop has 30 feet of vertical section in the upper one-quarter of the Sugarloaf Arkose about 600 feet below the Deerfield Basalt.

How to get there. - If you are going south on I-91, take exit 25, or going north take exit 24, and proceed east on Route 116 for about a mile. The outcrop is just east of the traffic light on Route 116 in South Deerfield. You can park a few yards to the east of the road cut in a parking area intended for people who want to hike the trail to the summit, a substantial and steep climb. When examining the outcrop or taking photographs, **be careful not to back away from the outcrop** and be hit by fast-moving vehicles on Route 116.

Comments. - Several observations imply that a braided river deposited the redbeds. 1) The river had a gravel-sand bedload with lots of cobbles, up to 15 cm in size. 2) There were many shallow channels, some of which eroded down into an underlying channel fill. 3) The high ratio of channel fill to floodplain mudstone contrasts with meandering rivers where overbank mudstones are abundant.

An unusual feature is that although the redbeds are mostly fluvial channel sandstones with cross-beds, the cross-beds are difficult to identify. Elsewhere at Sugarloaf Mountain, similar strata have lots of cross-beds. The cross-beds are hard to see because the structures in the sands were severely disrupted, probably by *Scoyenia,* a feeding burrow likely made by beetle larvae.

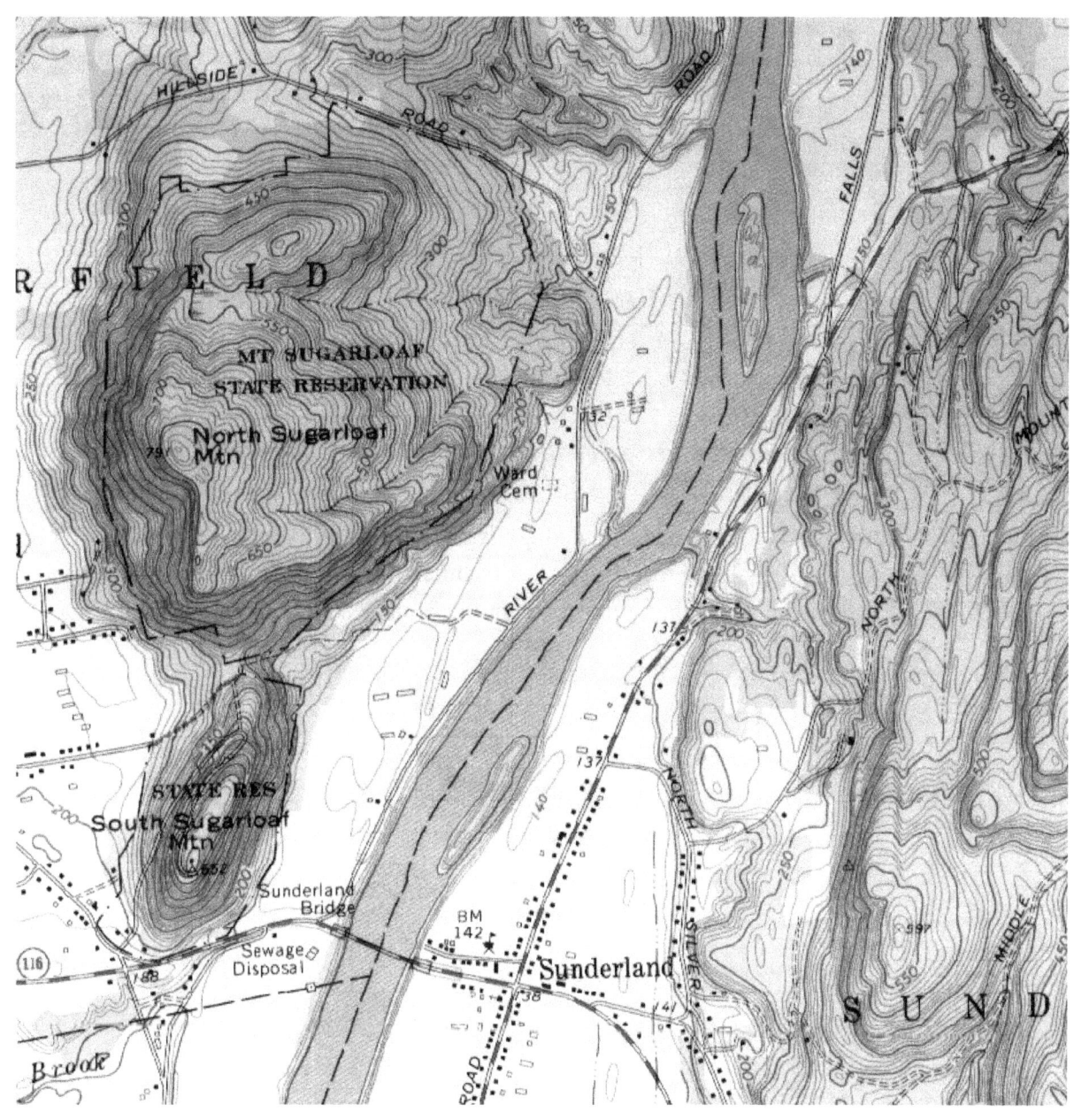

The outcrop is the road cut made by Route 116 at the south end of South Sugarloaf Mountain. Courtesy of the U.S. Geological Survey.

The outcrop is 6-m-high, 41-m-long and oriented 24 degrees from the average paleoflow direction of the river towards the 104-degree azimuth. River channels are evident to the left of the meter stick, with thin beds of flood-plain mudstones in the same area. Photo by Jim Dutcher.

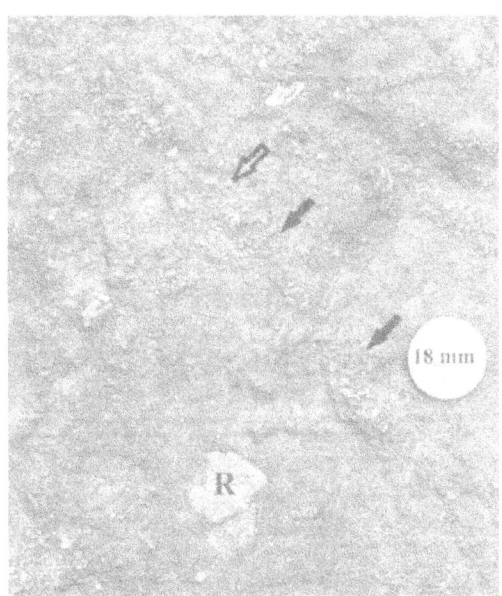

The fabric of the sandstones at this outcrop are severely disrupted, presumably by *Scoyenia* burrows, although identifiable burrows have not been found. This is a bedding-plane view of *Scoyenia* burrows in a pebbly sandstone in the Sugarloaf Arkose in Deerfield. The solid arrows point to back-fill lines that are concave to the upper left, the direction the beetle larva moved through the sand in search of food. The open arrow points to faint scratches on the outside of the burrow. R is a rock fragment.

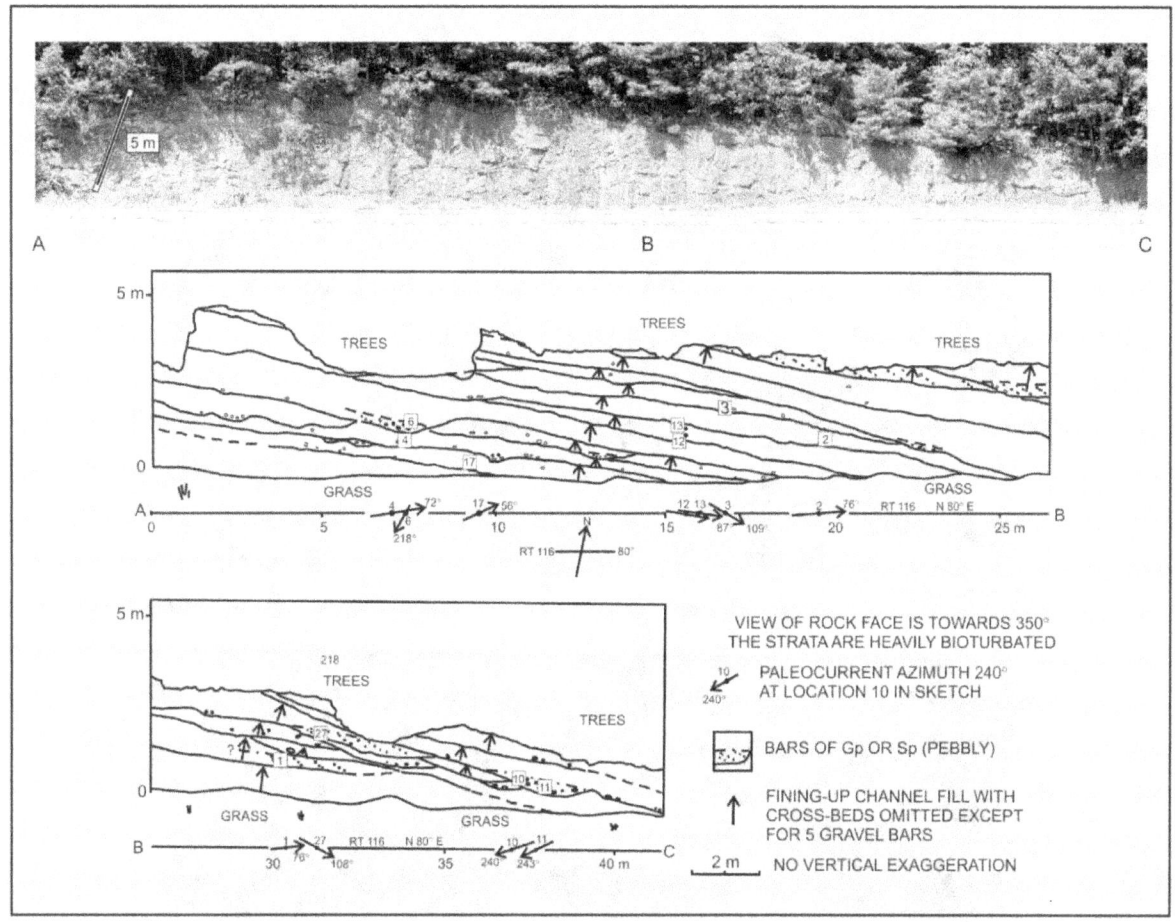

The rock face is subparallel to the direction of river flow, to the right. Cross-beds present in the upper right corner of the outcrop and ripple-marked sandstone at various places show that the river flowed southeast. The lines in the sketch trace the contact between adjacent sandstone channel fills, which average about a meter in thickness. The grain size became smaller upwards in each channel as it filled with sand and water depth shallowed. As a channel laterally shifted to a new location, the channel deepened down slope, scouring about half a meter into the underlying channel sands. Segments of gravel bars are in the deeper parts of some of the channels as at 7, 13, 26, 33, and 38 m on the horizontal scale below the sketch. From Hubert and Dutcher, 2010.

If you walk into the woods behind the parking area on Route 116 towards the rock outcrop, you can find ventifact pebbles lying on the ground, weathered out of the outcrop. The details of features seen on ventifacts are given in the description of the alluvial-fan deposit at Gill. Wind-blown sand sculptured grooves and ridges parallel to the direction of the paleowind. The facets of most of the ventifacts are polished.

Written in stone

Jim Dutcher, a geology professor at Holyoke Community College, and John Huburt, a geo-sciences professor at the University of Massachusetts, take measurements at a sandstone and conglomerate rock formation at the base of Mount Sugarloaf in South Deerfield on Monday.

A passing reporter from the *Amherst Bulletin* stopped to snap this photo published on August 27, 2004. Jim Dutcher and I are reading the strike and plunge of the crests of ripple marks in a sandstone at the Route 116 outcrop. River flow is towards the viewer.

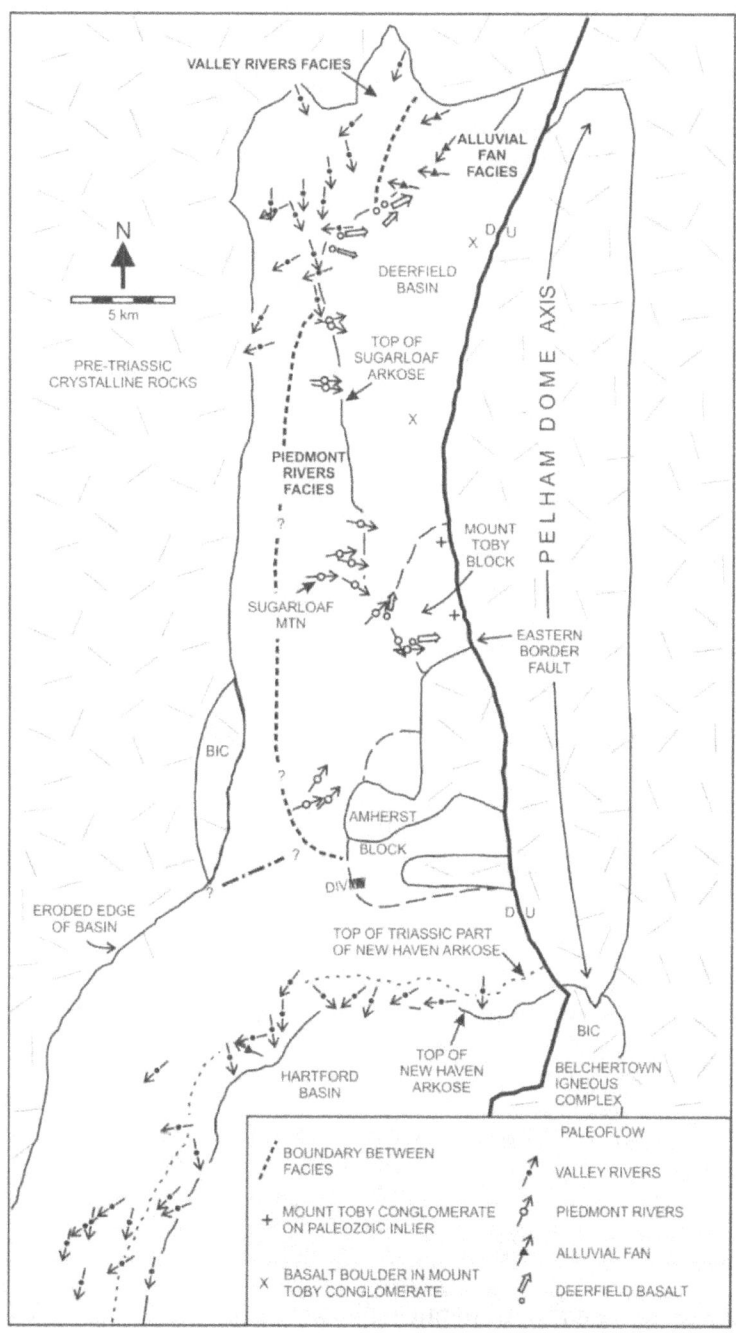

Paleocurrent directions for outcrops of fluvial redbeds in the Sugarloaf Arkose in the Deerfield basin and the equivalent New Haven Arkose in the northern end of the Hartford basin. Piedmont rivers refer to the bajada. The Deerfield basalt lavas flowed to the east down into the subsiding basin. The basalt boulders in the Mount Toby Conglomerate show that the lavas covered the terrain around the basin. From Hubert *et al.*, 2008.

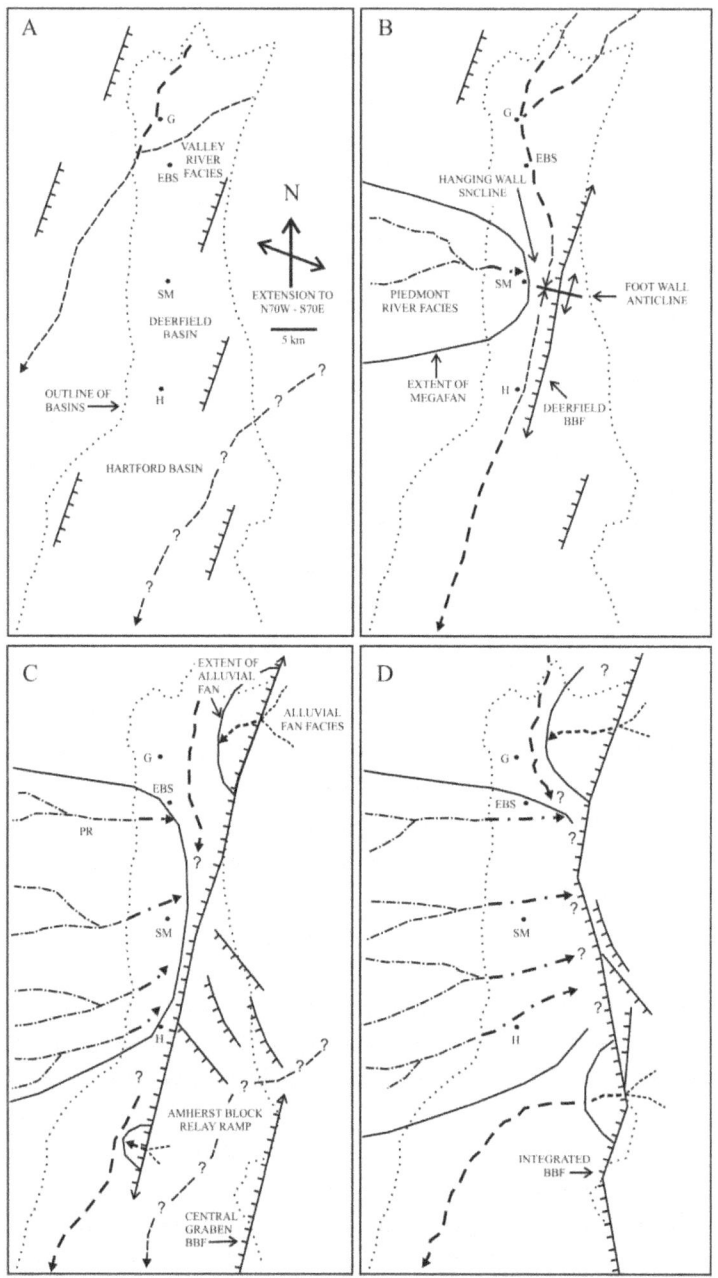

The diagrams show the tectonic and sedimentary history of the Sugarloaf Arkose. Shown are fault scarps, river channels, and fans. Omitted are abandoned features and channel tributaries. River channels are darker where outcrop data are available. A) Early basin sag before the border fault formed. B) The early border fault and the fan have formed. C) The fan and border fault enlarge. D) An alluvial fan develops in the northeast corner of the basin and the border faults in the Deerfield and Hartford basins link up. Locations are Greenfield (G), Eagle Brook School (EBS), Sugarloaf Mountain (SM), and Hatfield (H). From Hubert *et al.*, 2008.

The outcrop is the typical brownish red of the Mesozoic strata of the Connecticut Valley. Originally the sands and muds were yellow-brown due to limonite (hydrated iron-oxides) soil stains on the surface of the grains. The brownish red formed after deposition largely by the spontaneous dehydration over hundreds of thousands of years of the limonite to hematite (anhydrous ferric-iron oxide). A smaller amount of hematite pigment formed by pervasive post-depositional intrastratal dissolution of iron-bearing accessory minerals grains such as hornblende, augite, and biotite. The iron from the dissolved minerals goes into solution in the water between the sand grains, where it precipitated as hematite.

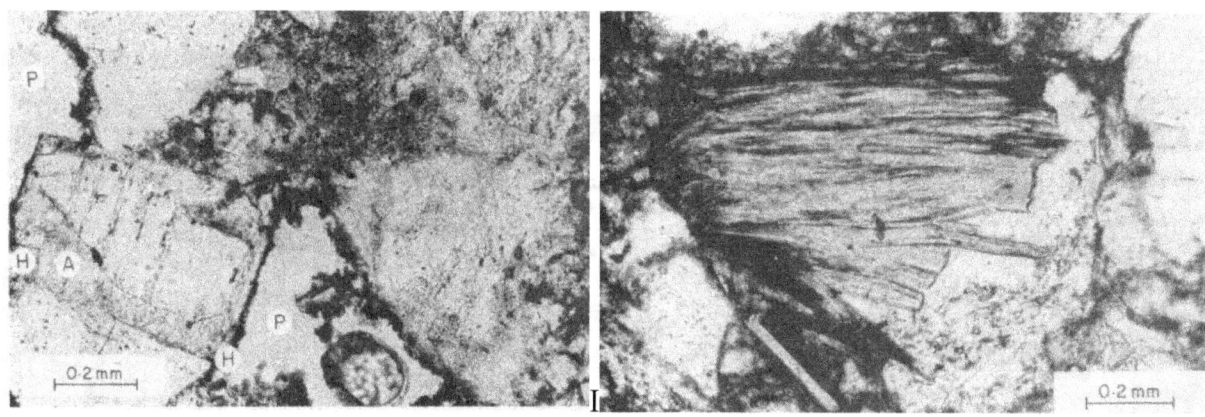

A thin section is a 30-micron-thick slice of a rock mounted on a glass slide and studied under a microscope. In the thin section on the left of a brownish-red sandstone, hematite pigment (H) was precipitated on albite overgrowths (A) that rim plagioclase grains in intergranular pores (P). Hematite-stained clay (dark) coats the grains beneath the albite overgrowths. On the right is a thin section of brownish-red sandstone where hematite (dark) has formed due to the dissolution of iron from biotite (center). Hematite was precipitated in the nearby intergranular openings. From Hubert and Reed (1978).

8. Poet's Seat Tower, Greenfield

What to see. - The summit provides a sweeping view of the western edge of the Deerfield basin and the Berkshire foothills. The tower is at an elevation of 492 feet, the highest point of Rocky Mountain Park.

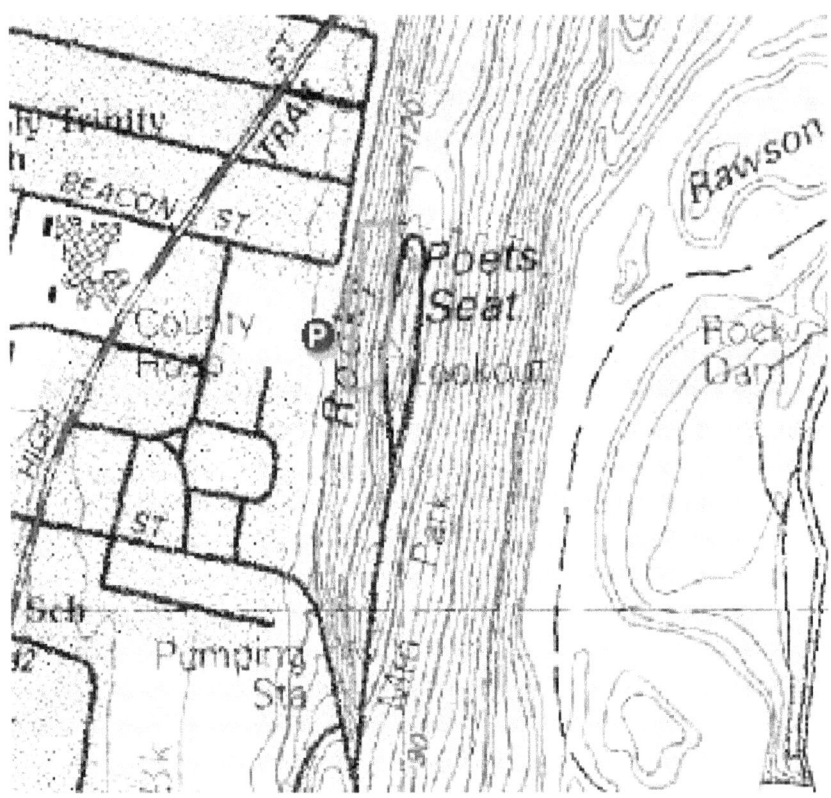

Beacon Field (P) provides parking in addition to that at the tower. Courtesy of Franklins.com. and the U.S. Geological Survey.

How to get there. - Once you are in Greenfield, the tower is visible on the high cliff of basalt to the east. The tower is just east of the County Hospital. The map shows the access road to the tower, where parking is available. The road to the tower is only open in the summer. If you prefer to walk up to the tower, a small parking area is available at the junction of Mountain Road and Parkway Street.

Comments. - In 1870, the Greenfield Rural Club constructed a wooden tower and an access road. The present tower was built of Turners Falls sandstone in 1912. The name is for the many poets who admired the spectacular view, particularly the Greenfield poet Frederick Goddard Tuckerman known for his sonnets in the Romantic literary movement. The tower was refurbished in 1977.

Poet's Seat Tower. Courtesy of Encyclopedia Britannica.

Two views of the tower. Courtesy of Timothy Clough on Flickr, the creative commons.

This engraving of Frederick Goddard Tuckerman (1821-1873) is the only known surviving image of the poet. Wikipedia.

    Here are the opening lines of "How oft in schoolboy-days" from Tuckerman's *Poems* (1860), courtesy of Wikiquote.

> How oft in schoolboy-days, from the school's sway
> Have I run forth to Nature as to a friend, —
> With some pretext of o'erwrought sight, to spend
> My school-time in green meadows far away!
> Careless of summoning bell, or clocks that strike,
> I marked with flowers the minutes of my day.

Looking to the west to Greenfield, Beacon Field is on the lower left. The 150-foot-thick, lower Jurassic Deerfield Basalt is in the lower right. In the distance is the junction of the western margin of the Deerfield basin and foothills of the Berkshires. The similar elevations of the hills record the Miocene-age plain formed by fluvial erosion, named a peneplain. On clear days, to the right out of the photo, you can see Mount Monadnock in NH, a resistant hill in the peneplain. Glacial Lake Hitchcock covered the valley floor with Rocky Mountain an island in the lake. Courtesy of Yankee Magazine.

9. Fluvial Redbeds, Country Club Road, Greenfield

What to see. - This long outcrop provides a cross section through fluvial redbeds in the late Triassic Sugarloaf Arkose. Exposed are channel forms, channel-fill sequences, trough cross-bed sets, mm-scale burrows in overbank mudstones, and a mold of a large tree.

How to get there. – Going south or north on I-91 take Exit 26 and at the rotary circle go east on Route 2A which turns into Main Street in Greenfield. Turn left (north) on Chapman Street and then right on Silver Street. Almost immediately you turn left onto Country Club Road (Swamp Road). Drive north to the outcrop, which is on the east side of the road just after the Interstate 91 overpass bridge and before reaching the Greenfield Country Club.

Note that the sun sets early in the fall after daylight saving ends.

Comments. - The numerous channel forms combined with minor mudstones show that river was braided rather than meandering. A meandering river deposits more mud on the floodplain than a braided river.

The map of cross-bed flow directions for the entire Sugarloaf Arkose at this stratigraphic level within the formation shows that the river flowed towards the south and southwest from highlands located to the north and east. The southeast flow here is part of the natural variability of river flow. Most of the Sugarloaf Arkose accumulated during the early stage of basin sag before the eastern border fault was established.

The two panels for Country Club Road overlap at the river channel on the right-hand end of the top panel. Meter stick for scale.

Grooves and scours on the base of the channel located near the fire plug present near the right-hand end of the top panel. The curved ends of the scours are the up-flow end, showing that flow was down to the right into the outcrop.

At the channel near the fire plug, river flow was southeast into the outcrop. 1) Transverse cross section of trough cross-bed set. 2) Channel sandstone where the trough cross-bed sets become finer-grained and smaller upward in the channel. 3) River bank eroded by channel flow. 4) Scour and groove marks on the bottom of the channel sandstone. 5) Cross-bedded pebbly sandstone of a bar that built from the bank into the channel. 6) Floodplain sandy mudstone pervasively bioturbated by elongate circular burrows a few mm in diameter. From Hubert *et al.*, 2008.

Before the late Triassic river switched to the location of this outcrop on Country Club Road, the river gradient had been lowered by a combination of the channel filling with pebbly sand and deposition of sediment on the adjacent floodplain. Seeking a steeper gradient, the river abruptly switched it channel to the current location, during which the flow eroded the channel into the underlying sediments. Over time, the new empty channel filled with pebbly sand while progressively becoming shallower. This process of channel shifting, deposition, lower gradient, and channel shifting was repeated many times as fluvial strata accumulated in the Deerfield Basin.

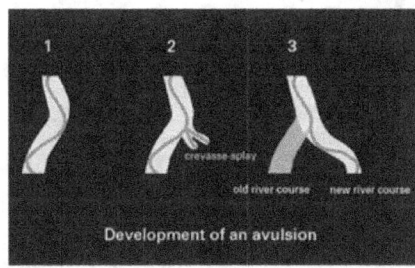

Channel shifting. Courtesy of Dept. of Physical Geography, Utrecht Univ., The Netherlands.

These grooves and scour marks are on the underside of the channel shown in the previous photo. A 10-cm wide scour extends from center left to lower right with grooves within it.

Grooves are linear marks cut into the river sand by pebbles dragged along in the current flow, in this case from upper left to lower right. The grooves are filled by the deposition of pebbly sand. Vortices in the turbulent river flow can erode elongate scours in the river bed, which also are filled by sand. The rounded end of a scour is up current and the open end is down current.

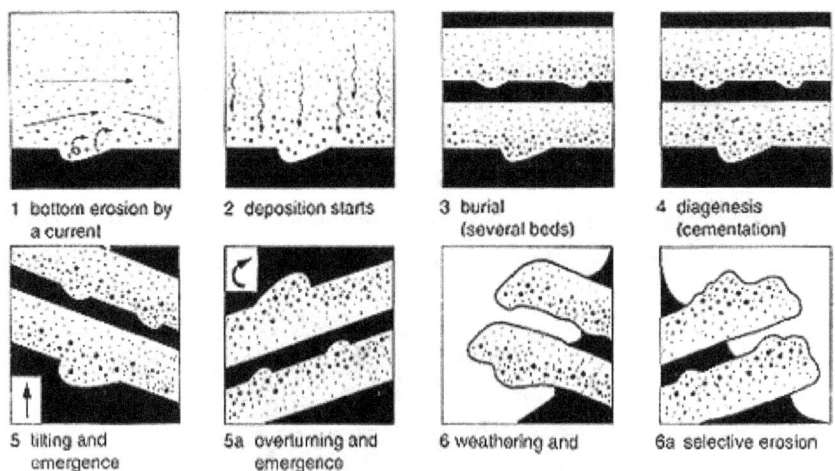

Grooves and scour marks are examples of sole marks observed on the underside of sandstone beds. The diagrams show how sole marks form. Courtesy of Columbia.edu.

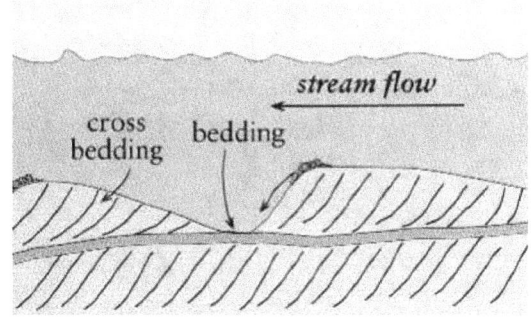

Downriver migration of subaqueous dunes produces cross-beds. Sand grains bounce and roll up the back side of a dune to the crest where they accumulate until numerous enough to avalanche down the front of the dune, forming a cross-bed. Courtesy of Kyle Brooks' blogspot.

Large river dunes with curved, cuspate crestlines are exposed after a drop in river level in Louisiana. Superimposed ripples developed in shallow water as the water level fell. Courtesy of seddepseq.co.uk

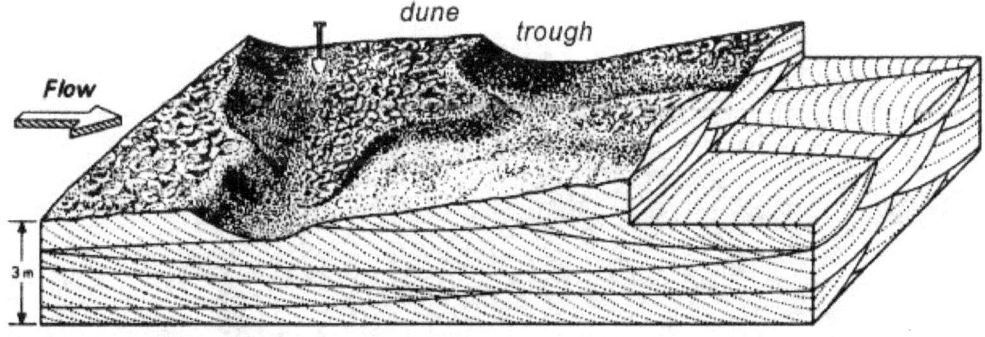

Trough cross-bed sets are made by downriver migration of subaqueous dunes with cuspate crestlines, which create a trough with scooped-out pits. As the dune advances, each pit forms a curved bounding surface of a cross-bed set, which in transverse section is trough shaped. Courtesy of geocaching.com. Original figure from Harms *et al.*, 1975.

Top: Cross-bed set with planar rather than curved bounding surfaces at Country Club Road inferred to be a straight-crested sand bar. Middle: Straight-crested sand bar in the South Platte River at Sterling, CO. I am digging a cross section through the slip face of the bar. Bottom: Planar cross-bed set exposed in cross section.

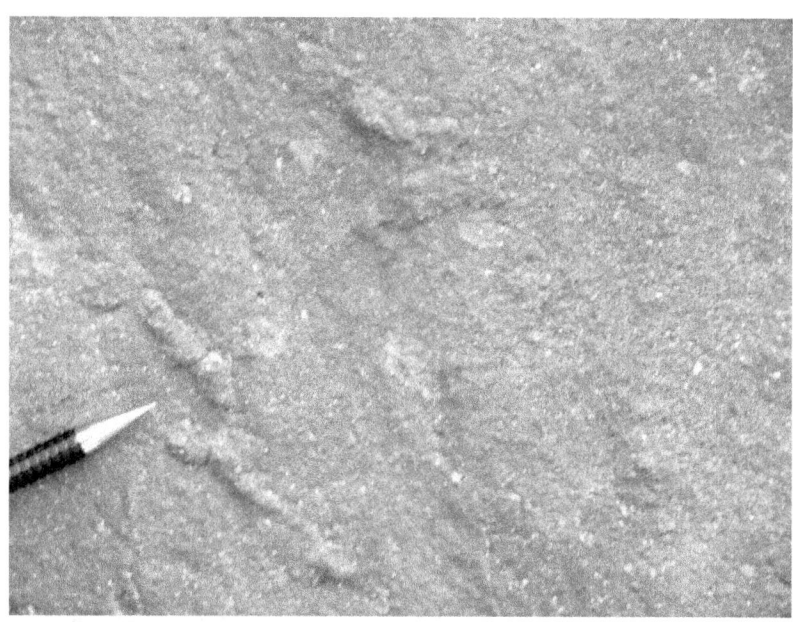

Scoyenia burrows on a bedding-plane at Country Club Road have curved back-fill lines concave to the upper left, the direction the beetle larvae borrowed through the sand in search of food. The beetles lived along a braided river in a monsoonal climate and may have had a life cycle similar to modern beetles. At the start of the dry season, the beetles flew over the river bed, landing to lay eggs in the moist, nutrient-rich sands and muds. As the dry season intensified, the larvae emerged from the eggs in the vadose zone above the water table and below the river bed. In the sands and muds, the larvae fed on decaying plant debris, microbes, and other nutrients with less chance of drowning by a rise in river level. The larvae formed pupae, passed through several growths stages, emerging as adults at the start of the rainy season.

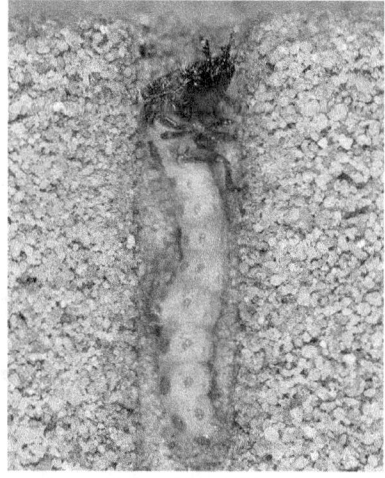

Northern dune tiger beetle larva in burrow. Courtesy of arkive.org and naturepl,com

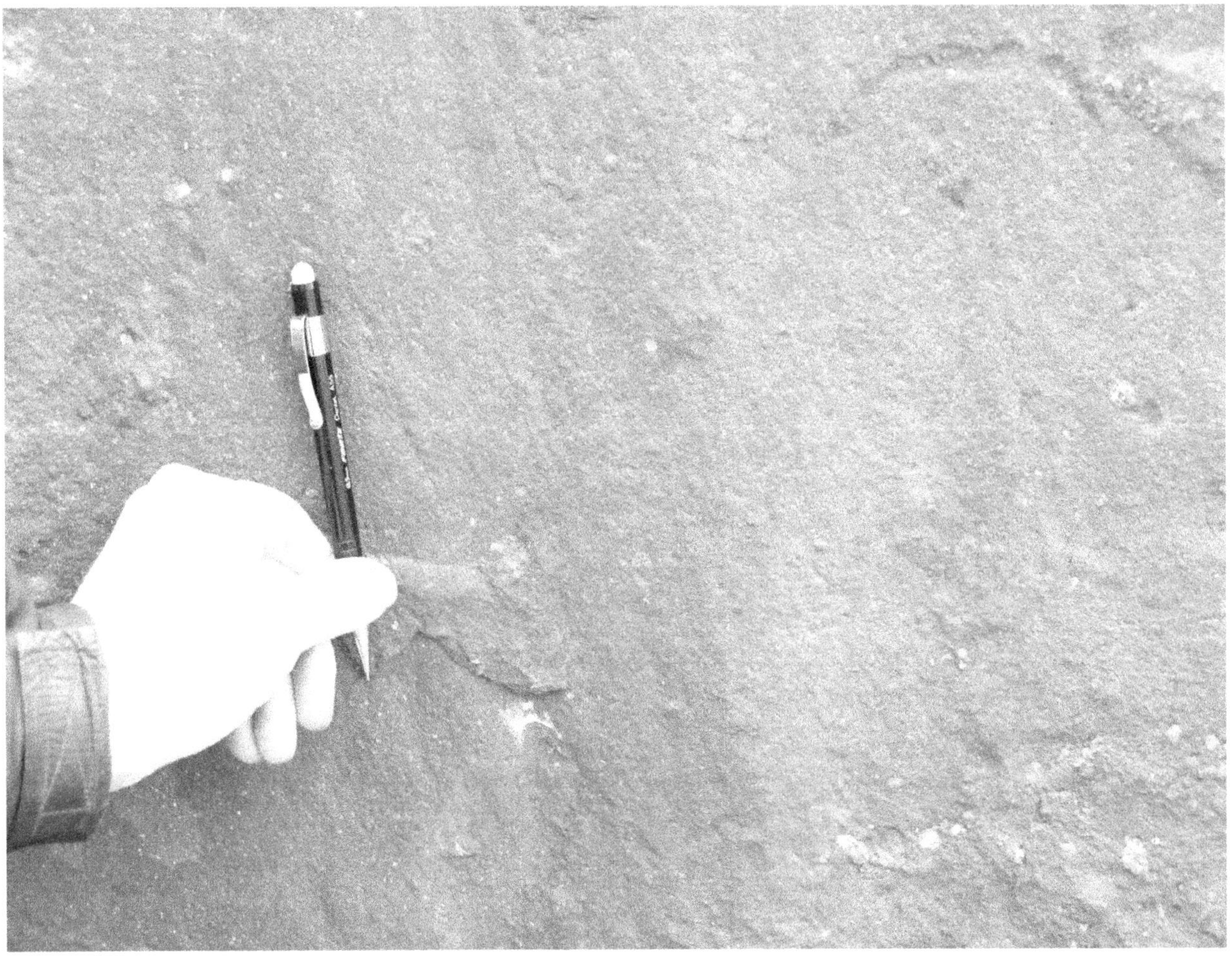

The grain lineation in a sandstone at Country Club Road is parallel to the pencil. The specific direction is chosen to match the paleoflow of nearby cross-beds and ripple marks.

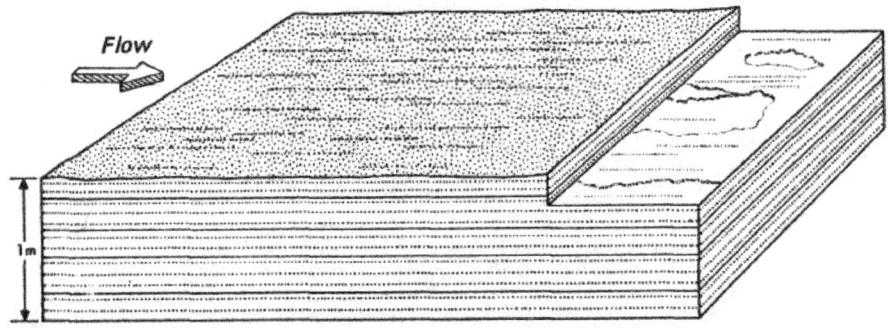

In grain lineation, the long axes of the quartz and other grains are aligned with the current and are imbricated, dipping up current. Courtesy of uga.edu and Harms *et al.*, 1982.

Mold of a tree trunk filled with muddy sand is present in channel sandstone near the I-91 overpass. River flow was into the outcrop. Dark-colored vertical root traces are to the left of the tree trunk. Plant roots thoroughly obliterated the depositional structures, such as cross-beds.

A braided river with numerous channels and sand bars. Son-Kul River in Kyrgyzstan. Courtesy of the University of Oregon.

The Susquehanna River floods Johnson City, NY, depositing a layer of mud as the water level recedes. The mud commonly contains 75 to 90 percent water which is expelled on compaction to mudstone. Courtesy of the NY Times.

10. Alluvial Fan, West Gill Road, Gill

What to see. - This outcrop is an alluvial-fan sequence built by flash floods from highlands in the northeastern corner of the Deerfield basin. The fan is near the top of the Sugarloaf Arkose, several km from the border fault of the basin.

How to get there. – At the bridge over the Connecticut River between Turners Falls and Gill take Main Road northeast for one-half mile to the center of Gill, where you turn left (west) on to West Gill Road. On the east side of Gill center there is a large green and a small round sign that says Town of Gill. Drive one mile to the outcrop, which is on the east side of the road.

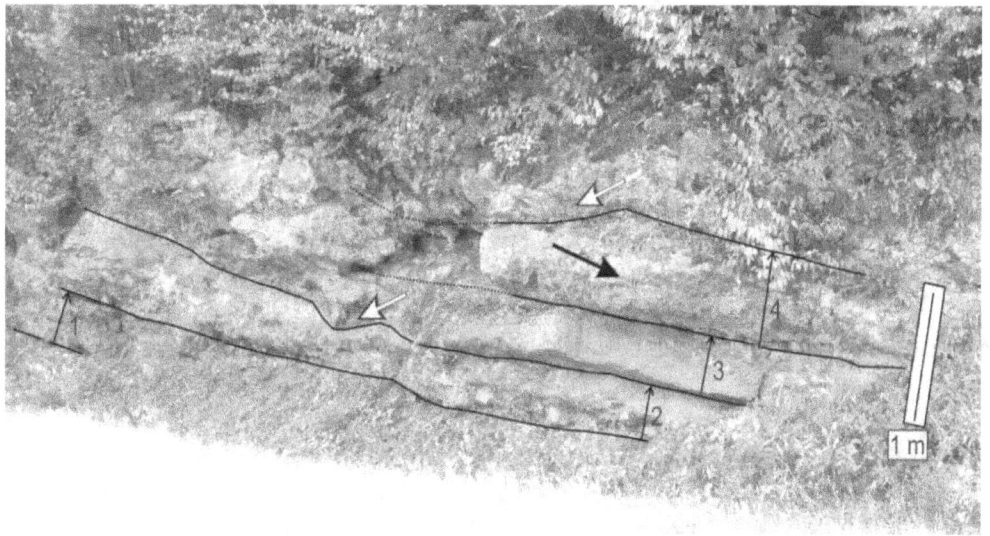

The four numbered beds are flood units of pebbly sandstone where the grains size fines upwards through the bed. At the base of the second bed are sub-angular to rounded, light-colored boulders up to 37 cm in size that rolled along in the shallow flood down the fan. The open arrows point to scour channels at the base of two beds. The solid arrow points to antidune cross-beds that dip to the left; flow was to the right. From Hubert *et al.*, 2008.

Comments. - Each graded bed records a shallow flash flood that coursed down the lower part of the fan, rolling boulders. As the flows spread, thinned, and decelerated, they deposited graded beds of pebbly sand, some with up-flow dipping, antidune cross-beds. The fan was cone-shaped in plan view, and at this location was building to the northwest towards an azimuth of 295°. The base of each bed is erosional, including scour channels from 19-42 cm deep, oriented downslope, cut during floods.

Close view of flood unit 4, showing imbrication of the disc shaped stones. The dark line (lower right side of photo) sloping to the left is an antidune cross-bed. Flow was to the right. How imbrication forms is shown at the Roaring Brook stop and antidunes at Chard Pond. The meter stick has 10-cm intervals. Photo by Jim Dutcher.

Absent are debris-flow beds recognizable by pebbles suspended in a sandy mud matrix. The reason is the scarcity of mud available from weathering of the crystalline rocks in the source area. Fans with sedimentary rocks in the source area commonly have debris flows

Ventifacts are present at the outcrop, fashioned by sand and dust driven by the wind. Some of the ventifacts have a surface coating of dark-colored, desert varnish. The varnish is a thin layer of iron and manganese oxides mixed with wind-blown clay particles that stuck to the rock. Wetting with dew and high temperatures in the sun aid the chemical processes that produce the varnish.

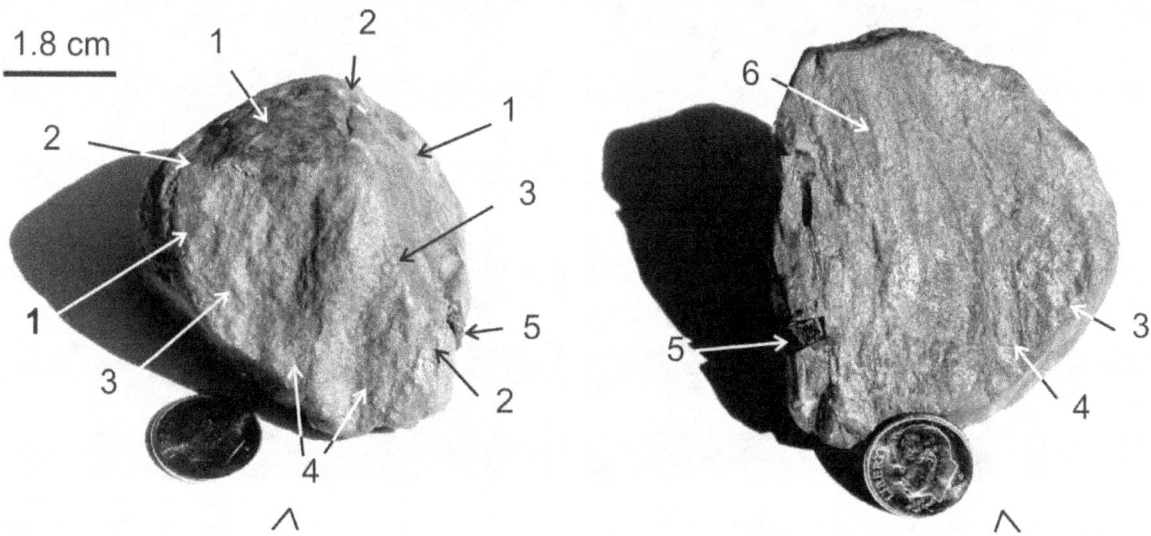

On the left is a view of the top and two sides of a ventifact from West Gill Road. Features with polish are 1) facet, 2) keel on the upper surface parallel to the dominant wind direction, 3) groove, and 4) pit. On the facet facing the dominant wind, the grooves run up the facet, parallel to the dominant wind. 5) A clay plug fills a missing piece of the ventifact. On the right is the base of the ventifact, which has less well-developed polish and grooves. 6) Thin layering of the metamorphic gneiss rock. The high polish indicates that blowing dust was important in forming the ventifact. The bottom of the ventifact is lightly polished, implying the stone rolled over or wind blew dust under the stone. From Hubert *et al*., 2008.

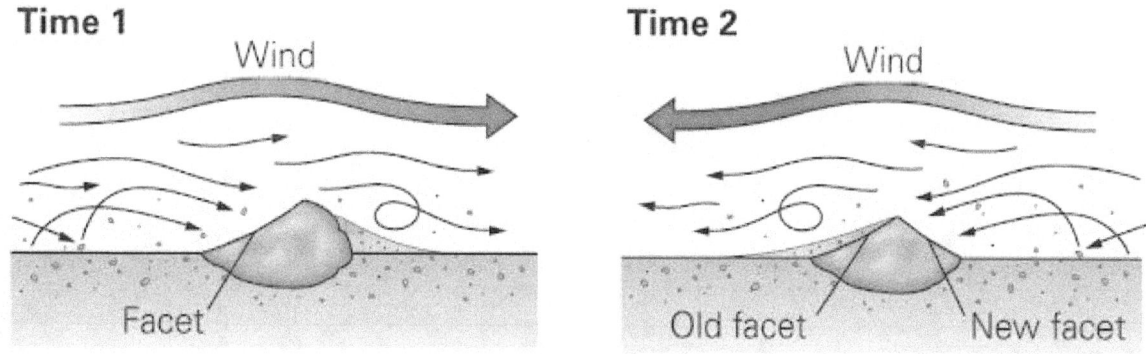

In time 1, windblown sand sculps a facet. In time 2, a change in the wind direction, which can be seasonal, forms another facet. Courtesy of geologylearn.blogspot.

The alluvial-fan deposits in Death Valley, California, are analogues for the fan at Gill. The areas between the light-colored, active channels are darker due to the growth of plants. The white areas on the playa are salt from evaporation of a temporary shallow lake. Courtesy of the U.S. Geologic Survey.

Alluvial fans along the fault-bounded eastern esarpment of the Deerfield basin. Courtesy of James Wessel.

11. Lacustrine Strata, Barton Cove Campground, Gill

What to see. – This location has exposures of lacustrine gray strata in the Turners Falls Formation. The mudstones have carbonate laminae that reflect precipitation of Mg-bearing calcite from the surface water during the annual dry season. The breccias and folds in the strata are of tectonic origin rather than slumps that slid down the margin of the lake.

The site is maintained by FirstLight Power in Northfield, MA, for the owner GDF Suez Energy North America. The campground is open Memorial Day weekend through Labor Day. Each campsite has a picnic table, grill and campfire ring. The nature trails and parking are open all year. Also available are rental canoes and kayaks. For many years a pair of bald eagles with some mate changes nested on an island in the cove, where they could be observed on the web by an eagle-cam.

Warm clothing is advisable in the late fall when the wind blows off the river.

How to get there. – The campground is at 82 French King Highway, Gill. From north or south on I-91, take exit 27 and drive east on Route 2 for 3.4 miles. Turn right into the campground at the second Barton Cove sign. From the east on Route 2, cross over the Connecticut River on the French King Bridge and proceed 2.3 miles to the entrance to the campground, which is at the base of a steep hill on the left side of the road.

Comments. – From the parking area, walk up the dirt road for about 50 yards and take the trail down slope to the right to a wooden platform overlook where you can observe the lacustrine laminae, breccia, and folds.

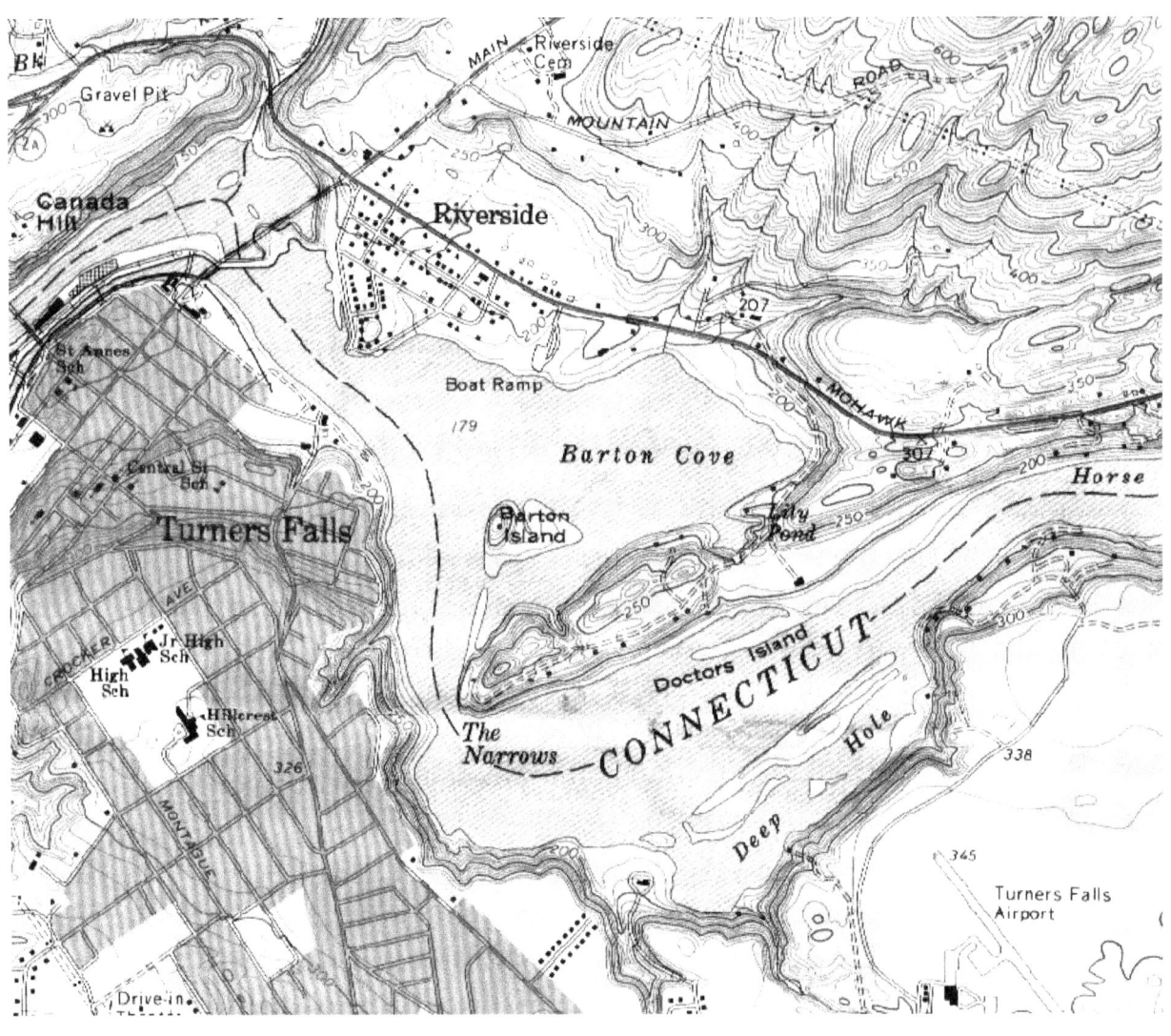

Barton Cove Campground is on the peninsula that juts out into the Connecticut River. Courtesy of the U.S. Geological Survey.

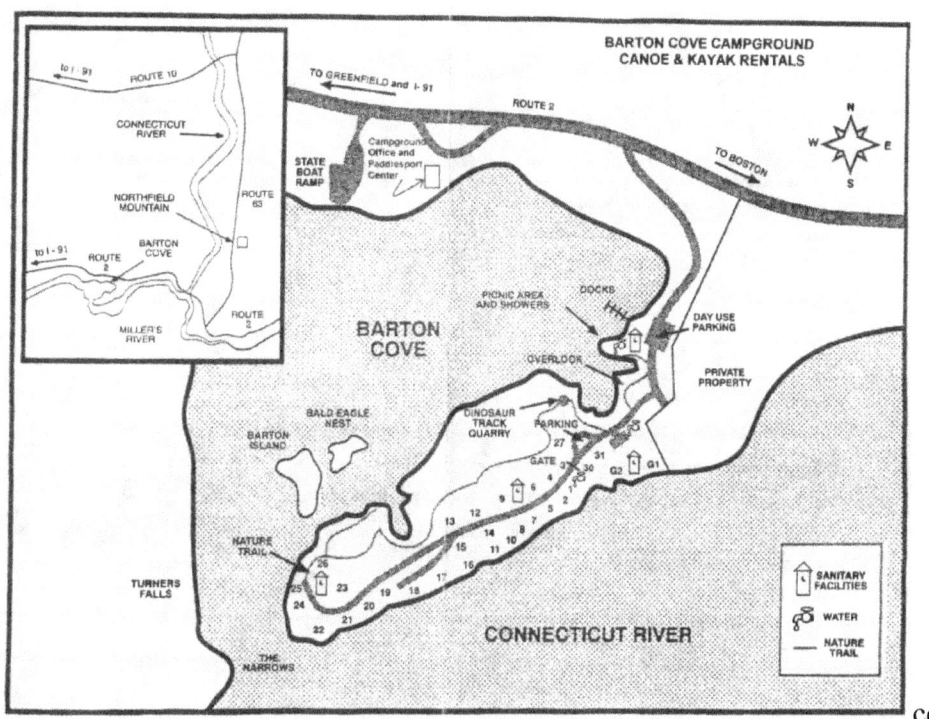

Barton Cove Campground. Courtesy of Northampton Chamber of Commerce.

The breccia (left center) is shown below. Courtesy of Planetward.com.

The light laminae in the blocks of laminated dolomitic mudstone are dolomite, the Ca/Mg carbonate mineral whereas the dark are clay mineral. The lake was alkaline with substantial dissolved Ca, Mg, and bicarbonate ions. During the annual dry season, intense subtropical evaporation caused concentration of the ions in the surface water, while photosynthesis increased the alkalinity to above a pH of 8.2, causing precipitation of tiny crystals of Mg-bearing calcite (calcium carbonate). After burial, the calcite altered to dolomite in the Mg-rich pore water around the crystals. Courtesy of Kurt Hollocher.

The laminae are not borrowed by worms or other animals living in or on the lake floor because the lake water was thermally stratified with an oxygen-rich upper layer mixed by wind storms over a lower oxygen-deficient layer not mixed by the wind. Fish and other animals lived in the upper layer whereas the lower layer did not support life.

During the summer dry season, mud-size (< 62.5 μm) calcite crystals are precipitated by evaporation of the surface water of a lake in Turkey. Courtesy of climatica.org.uk.

- **Mixolimnion-** the zone that mixes completely at least once a year
- **Chemocline-** The intermediate layer, where there is a sudden change in density at the upper edge of bottom layer accumulating salts or dissolved organic matter.
- **Monimolimnion-** the non-mixing bottom layer

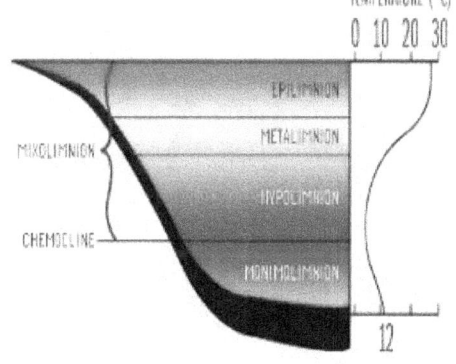

The laminae in a meromictic lake are preserved without burrows because borrowing animals such as worms cannot live in the oxygen-deficient bottom layer (monimolimnion). Courtesy of Gerome Rosario and slideshare.net.

The angular fragments were lithified before brecciation, indicating the breccia is not a slump deposit. Photo by Jim Dutcher.

Fold in laminated dolomitic mudstone located near the wood walkway. The rocks above the fold moved to the upper left whereas the rocks below the fold moved to the lower right, showing that the direction of motion was to the upper left. Photo by Jim Dutcher.

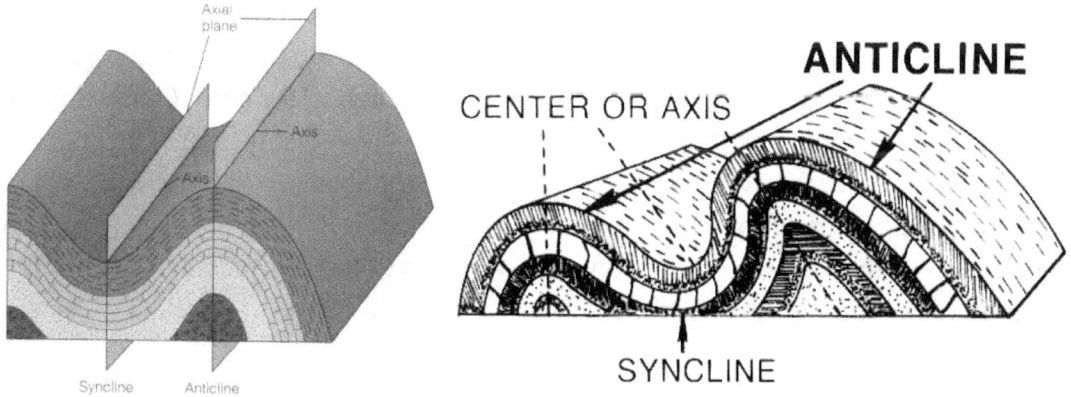

Axial plane and axis of a syncline and anticline. Wikipedia. Anticline and syncline. The direction of motion in the anticline is to the upper left. Courtesy of lumenlearnhng.com.

A large fold in dolomite mudstone is present near the walkway. For years, the question was whether the folds and breccia were caused by slumping of sediment layers down the edge of the lake, or due to tectonic stresses of later origin. The folds and breccia are now known to be of tectonic compressional origin as evidenced by the folds axes that are mostly oriented east-west while the axial planes dip to the south with the direction of motion to the north (Wise et al., 1992; Olsen et al., 1992). Paleocurrent readings in the playa redbeds show that the rift valley floor sloped to the southwest, not to the north. Also some folds pass transitionally into breccia masses. The deformation was caused during folding of the strata in the Deerfield basin into a synclinal trough with its northwest-southeast axis located south of Barton Cove. As the syncline was forming, the beds were compressed and squeezed to the north out of syncline (Olsen et al., 1992). Courtesy of Kurt Hollocher.

Plunge pool seen from the wooden walkway. Photo by Jim Dutcher.

From the air, the two river channels leading to the plunge pools are clearly visible, but hard to discern on the ground. For years bald eagles nested in a tall dead tree on the smaller of the two islands in the cove. Before the tree toppled in 2008, nearly three dozen bald eagles fledged from the island. The Turners Falls airport is at top left. Courtesy of Planetward.org.

During the stable phase of Lake Hitchcock, the Millers River built the Montague delta that extended across the lake basin, completely burying the bedrock peninsula at Barton Cove and the surrounding area (Brigham-Grette and Wise, 1988). The delta created a narrow, shallow lake basin in the north separated from a deeper, broader lake to the south. As Lake Hitchcock drained and the lake level fell, the Connecticut River cut into the Montague delta, removing the unconsolidated sand and gravel until the river became perched on the ledge at Barton Cove, creating spectacular waterfalls with plunge pools. Two abandoned river channels lead to the plunge pools. The waterfalls lasted for a few thousand years until abandoned some 10,200 years ago (Curran, 1999). The Connecticut River continued to erode the delta sands, finally reoccupying a pre-glacial Connecticut River channel located at the southeast end of the ridge.

In 1868, Alvah Crocker from Fitchburg, Massachusetts, founded Turners Falls as a planned industrial community, using the 41-foot drop of the Connecticut River at the "Great Falls" to generate cheap hydropower through the construction of a dam and canal. Before the dam was built, the plunge pools were known as the Lily Ponds. The nearby Hitchcock dinosaur-track quarry was named the Lily Pond quarry. The dam raised the river level upstream from the dam, drowning the ponds and creating a cove.

Professor Edward Hitchcock (1793-1864) of Amherst College is forever associated with the dinosaur tracks because he amassed a huge collection and named many species. There is, however, the oft-told story of the unfortunate spectacle of Hitchcock and Dr. James Deane (1801-1858) both claiming to have discovered the dinosaur tracks when actually Dexter Marsh (1806-1853) did discover them. The tale is told by Paul Jenkins in his book, *The Conservative Rebel: a Social History of Greenfield, Massachusetts.*

"Dexter Marsh, a day laborer and night janitor at Town Hall, had discovered in 1835 what he guessed to be bird tracks in sandstone slabs he was laying as sidewalk along Bank Row. He went back to the source the red banks of the Connecticut at Riverside, and began building a personal collection of fossil specimens, convinced he was on to something. When he showed his finds to James Deane, a local physician who doubled as an amateur scientist, Deane sent plaster casts to an acquaintance of his, Edward Hitchcock of Amherst College, a world authority on such subjects"

"Skeletons of dinosaurs had been discovered for the first time some fifteen years earlier in England. But no one had ever thought to connect these rock prints with such unimaginable things. Hitchcock speculated that Marsh and Deane were right in attributing these tracks to prehistoric birds. The discovery was important enough to draw Thomas Huxley from London to view the slabs. In the scientific community Hitchcock went on to receive credit for their discovery and identification. Deane wanted the credit for himself. Marsh seems not to have cared, though he was elected to the national Academy of Natural Science. Only afterwards was it realized that the prints had been made by dinosaurs walking on their hind legs."

Dexter Marsh discovered and recognized the significance of the dinosaur footprints. Marsh had a "cabinet" (museum) in a hall in his self-built home in Greenfield where he displayed his personal collection of reptile footprints and trackways, fossil fish, Indian antiquities, and "curios." Never formally educated, he was elected a member of the American Association for the Advancement of Science (1846), and in 1852 the Lyceum of Natural History in New York and the Academy of Natural Science in Philadelphia. Photo by Ed Gregory. Courtesy of the Jurassic Road Show.

Hitchcock's reptile footprint quarry was active in the 1800s. During floods, mud and sand accumulated on the playa surface, now seen as extensive, smooth bedding planes. Mudcracks and ripple marks are common. The dolomite nodules precipitated by post-depositional evaporation of Ca-Mg-rich pore water in the muds and sands. Photo by Jim Dutcher.

The display slab located near the top of the stairs has dinosaur tracks and mudcracks, typical of a playa setting.

One of the tracks on the display slab. Scale intervals are 10 cm. Photos by Jim Dutcher.

12. Lacustrine Delta and Alluvial-fan, Chard Pond, Sunderland

What to see. - A long cliff exposes a lacustrine delta sequence in the Turners falls Formation and the overlying alluvial-fan deposits of the Mount Toby Conglomerate. The deltaic sandstone has spectacular pillow structures.

The outcrop is in the Mount Toby State Forest and open to the public. Do not hammer, deface or collect samples of the pillow structures or any other part of the outcrop. Geoscience departments regularly visit the outcrop.

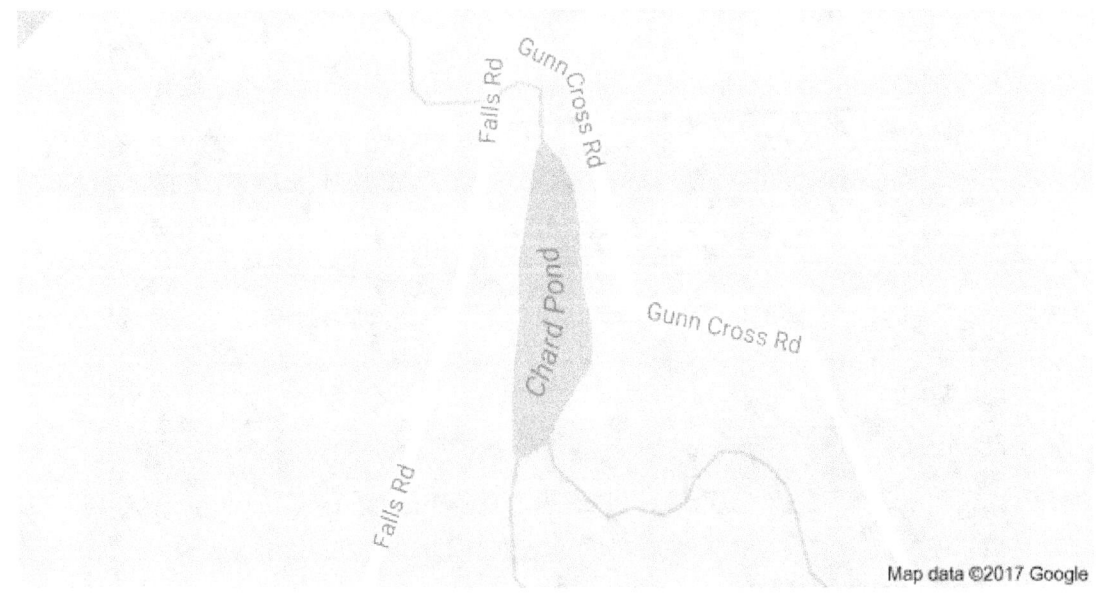

The outcrop is on the east side of Gunn Cross Road at the intersection with Falls Road. Courtesy of Google.

How to get there. - Starting at the intersection of Routes 116 and 47 in Sunderland center, take Route 47 to the north. Exit diagonally off to the left onto Falls Road. Proceed down Falls Road parallel to the Connecticut River to Chard Pond, which is on the right. You can park beside the gravel Gunn Cross Road that goes uphill to the east between the cliff and Chard Pond.

The dashed line marks the contact between the deltaic lacustrine strata and the overlying alluvial-fan pebbly sandstone and conglomerate. The delta foreset beds (F) dip at a low angle to the northeast (left). A break in grain size (arrow in lower center) is present between the finer-grained lower-slope and coarser-grained upper-slope deposits of the delta. Pieces of lithified lacustrine sandstone in the lowest beds of conglomerate show that the lake dried up by evaporation before the fan covered the lacustrine strata. The conglomerate records shallow, flash floods that coursed down the lower part of the fan, depositing antidune laminae (A). The imbrication of the plate-shaped pebbles shows that the fan advanced to the northwest.

On leaving your vehicle, walk over to the contract (white line) between the Turners Falls Formation and the overlying Mount Toby Conglomerate.

If you walk to the north along the cliff from the previous view, you come to pillow structures that formed on the delta slope. The pillows are about 6 to 8 times as long as wide. Two geologists examine a pillow horizon and the underlying slide surface that slopes to the left.

Comments. - When you look at the outcrop, think back to the early Jurassic 200 million years ago when nothing you see about you was present. This location was in a subtropical rift valley located 25 degrees north of the equator in a semi-arid climate with seasonal rainfall. Several kilometers to the east, a border fault escarpment was present between the subsiding basin and mountains. Erosion in the mountains provided the gravel and sand to the fan. During prolonged wetter cycles, a lake lapped up onto the toe of the fan, fish swam in the lake, and dinosaurs roamed about. Over time, the lake dried up and the fan advanced over the surface of the lake strata.

Origin of the pillows. - The pillows formed by rapid collapse of the fabric of a layer of sand accompanied by expulsion of the intergranular pore water. The sequence starts when the river in flood stage transports medium and fine-grained sand to the delta, where the current spreads the sand over the delta platform and down the delta slope. The sand is rapidly deposited as horizontally laminated or cross-bedded layers. The fabric of the sand is a "house of cards" with about 60% porosity, helped by the 30 percent mica flakes. For comparison, sand bars in modern rivers have 32 to 52 percent porosity.

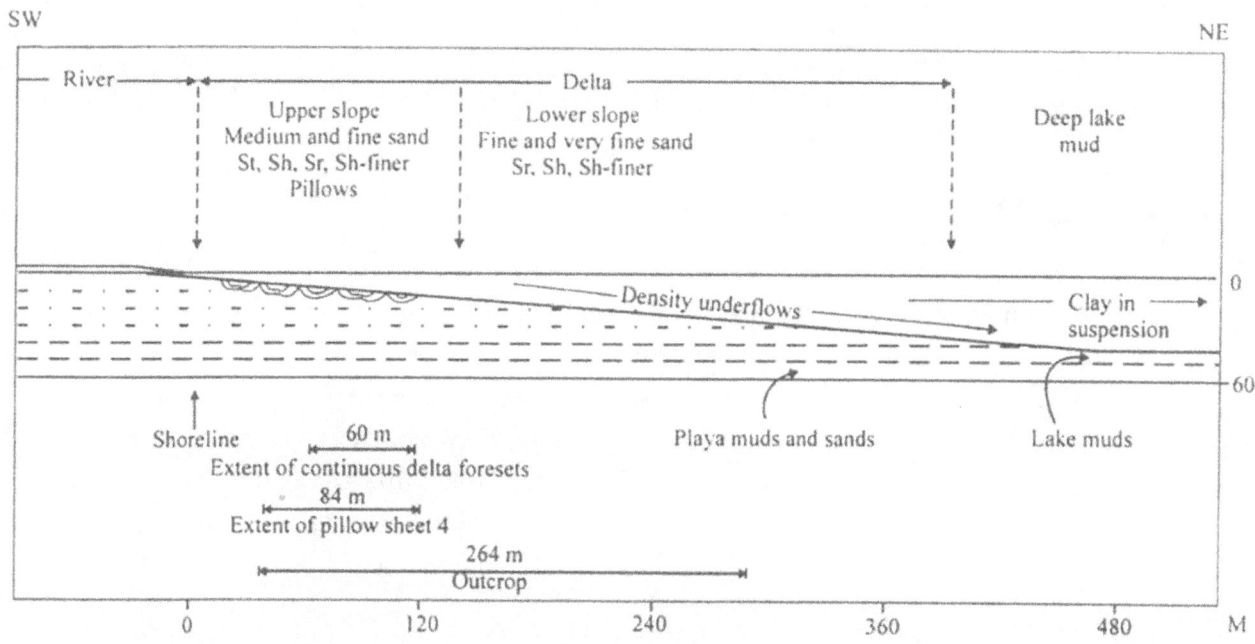

Cross section of the lake parallel to the direction of delta building to the northeast. St is trough cross-bedded sandstone; Sh is horizontally-laminated sandstone. The sequence of gray lake beds is underlain by playa redbeds exposed in the gully located across Falls Road. The vertical and horizontal scales are the same. From Hubert and Dutcher, 2005.

Three pillow horizons are P2, P3, and P4. You are looking about parallel to the direction of the pillow axes and the paleo-contours on the delta slope. Soon after formation of the pillows, currents beveled them, followed by deposition of sand. Pillow layer (P3) thickens downslope to the northwest (left) along a slide surface. The large solid arrow points to the contact between the muddy, fine-grained sandstone of the lower delta slope and the fine to medium-grained sandstone of the upper delta slope. The delta foreset beds (F) dip northwest. The meter stick has 10-cm divisions. From Hubert and Dutcher, 2005.

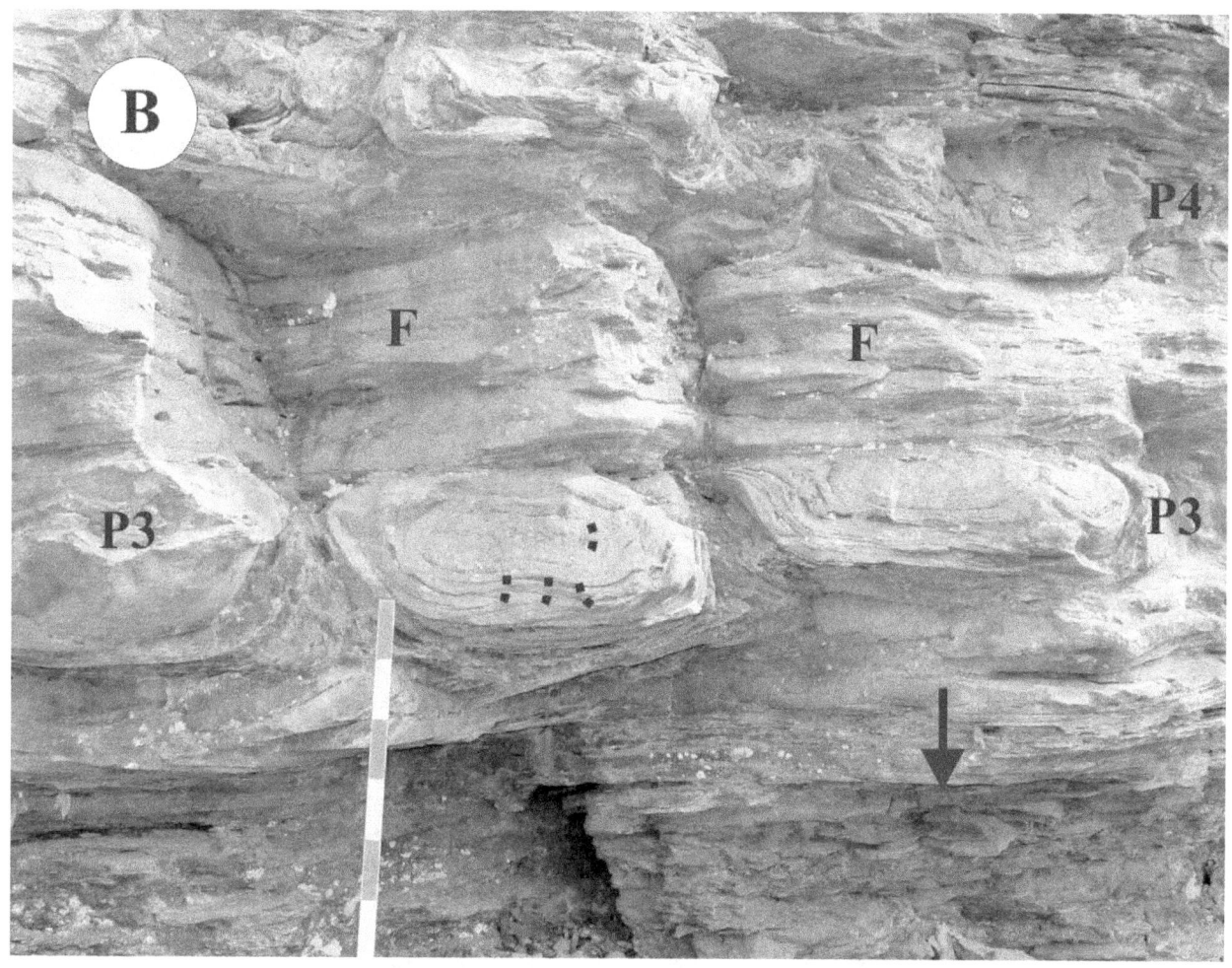

In pillow sheet 3 (P3), a sandstone pillow (right center) slid downslope over the edge of the next pillow. A fluidization dike (d) is present between the two pillows on the left. The dots on the center pillow mark water-escape pipes. The solid arrow points to the contact between the finer-grained lower and coarser-grained upper deposits of the delta slope. The meter stick has 10-cm intervals. From Hubert and Dutcher, 2005.

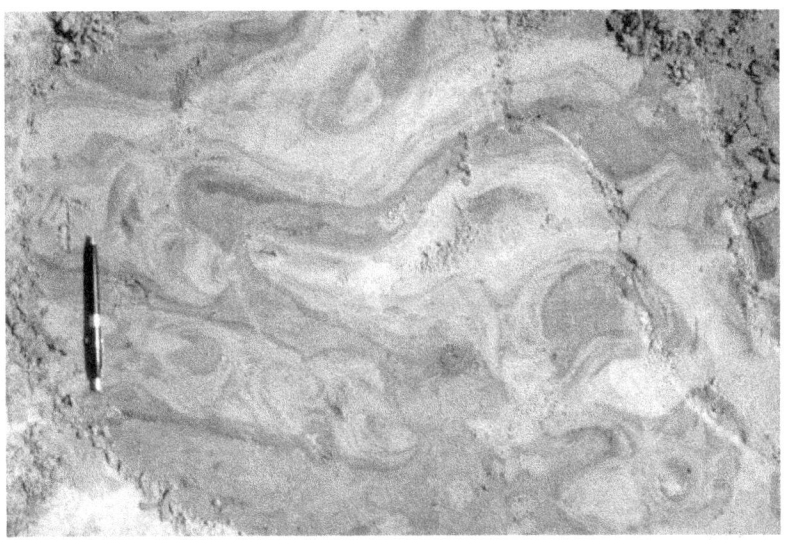

Similar pillows formed during deposition on a point bar of the Mississipi River near Baton Rouge, LA. A smoothed section through the point bar perpendicular to current flow has cross sections of the pillows that resulted from collapse of the integrity of the sand frabric during rapid deposition. The dark and light layers in the pillows are variations of grain-size due to fluctuations of the current velocity. The darker layers contain more mud. Below the pillows are "climbing ripples," which form during rapid accumulation of sand. As the ripples accumlated, they climbed to the right at a low angle from horizontal. The dark cross-laminae in the ripples contain organic matter. Gail Ashley did flume experiments at M. I. T. in the early 1990s that show a 20-cm-thick sequence of climbing ripples was deposited in 45 minutes from a dense underflow slowing down from 60-cm per second to zero. A smoothed, horizontal surface shows that the pillows are sinuous, elongate structures. The pen points down slope.

Cross section of a pillow that formed while deposition was proceeding. The sand sank forming an empty trough, which was filed by inclined cross-beds of sand transported from right to left. The cross-beds are not folded by the pillow. Strong currents then cut an erosion surface that bevels the trough and cross-beds. The pillow is located near the far southeast end of the cliff.

As a layer of sand rapidly thickens by continuing deposition, the fabric collapses and the grains move closer together, expelling the water up out of the sand. As the grains become progressively more closely packed, the density of the sand increases beyond that of the underlying water-saturated mud or sand, causing the sand to sink into the underlying layer. The sinking sand forms elongate pillows rather than bowls because of the tension created by the presence of the slope, with the pillow axes oriented across slope. In some of the collapsing sand layers, a sheet of pillows slid a few inches to several feet down slope on a slide surface.

The initial view of geologists was that an earthquake along the nearby border fault of the basin shook a surface layer of sand, causing the sand fabric to compact, expelling the pore water, accompanied by sinking of the sand layer into the underlying mud. However, the pillow sheets are syndepositional structures as evidenced by pillows where currents transported sand into the open troughs of sinking sand, depositing inclined laminae that are not bent. Then the pillows sheets were beveled by currents.

Flood layers (solid arrows) deposited on the upper slope of the delta fine upward from sand to mud (dark). Pipes and thin dikes (dots) show where water escaped upwards out of sand layers during compaction of the sand fabric. P4 is the forth pillow layer. The open-tip arrows point to the contact between the lower and upper-slope deposits. From Hubert and Dutcher, 2005.

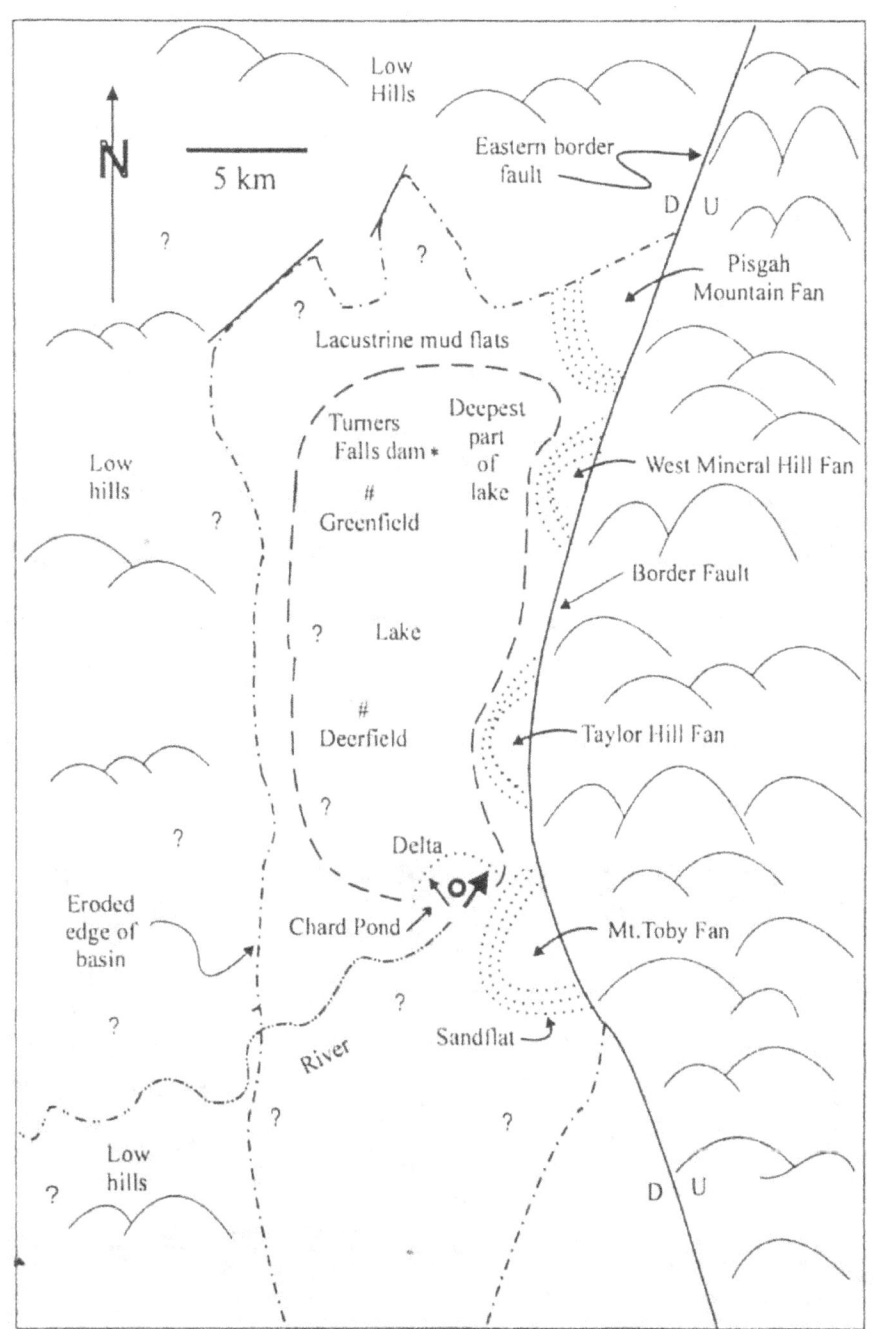

Paleogeography of the lake, which deepens to the east toward alluvial fans along the border fault. The arrows at Chard Pond show the direction of delta advance, which was mostly to the northeast. The river that built the delta came from the southwest. Areas without data are shown with question marks. From Hubert and Dutcher, 2005.

Similar to the situation at Chard Pond, the shallow, alkaline, sodium-rich Lake Manyara laps onto the toes of alluvial fans along the fault escarpment of the east African rift valley in Tanzania. Lake Manyara has a maximum depth of 12 feet so that during prolonged dry spells large areas of mud flats are exposed and animals leave their foot prints. Courtesy of alltimesafaris.com.

During a storm, a rapid, shallow stream emerges from the fault-bounded mountains at the apex of an alluvial fan in Death Valley, California, and travels down the light-colored channel, depositing pebbly sand that builds up the fan surface, reducing the slope gradient. After a time, the channel shifts to a steeper gradient elsewhere on the fan, building a 180-degree cone. The active channel is light colored because desert plants grow outside the channel. Courtesy of newgeology.us.

Walking a few feet to the right (southeast) from the original view of the outcrop takes you to an excellent view of the alluvial-fan strata. Warmly dressed field trippers watch co-leader Pete Panish point to the base of a flood layer in the alluvial-fan conglomerate while I carefully make my way down slope.

The flood deposits in the previous photo start with unit 1, which rests on the eroded surface of the lacustrine strata. The flood layers were deposited on the lower part of the fan by shallow sheet floods with antidunes. The individual layers tend to become finer-grained upward. Antidune cross-beds dip at a low-angle to the left, particularly evident in units 2 and 5. Compared to cross-beds in subaqueous dunes, antidune cross-beds are coarser-grained, thicker, and dip at a lower angle from horizontal.

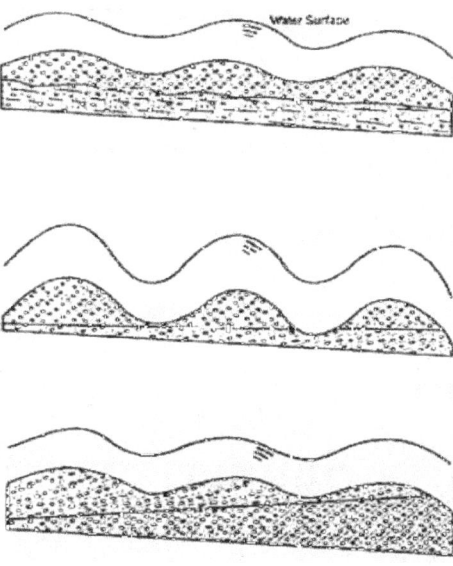

Formation of antidunes with river flow to the right. *Top*: start of antidune formation. The grains bounce and slide down river, landing on the upcurrent sides of the sand waves. As sand accumulates, the sand waves move in the up-river direction. At the same time, sand is removed from the downriver side of the antidunes so overall sand is moving down river. The sand waves and the sutface water waves are in phase and both move together up river. If a sand wave becomes over steepened, then the surface wave breaks in the up-flow direction. *Middle*: steepening of the surface water and sand waves. *Bottom*: antidune migration to the left with slight amplitude reduction while the cross-beds dip up flow. Courtesy of Terence C. Blair.

Antidunes in the fast-flowing Kennnetcook River in Nova Scotia. Flow is to the right, but antidunes are visibly migrating to the left, breaking in the up-flow direction. Courtesy of www.st.ca.

13. Mount Toby Conglomerate, Roaring Brook, Sunderland

What to see. – The Mount Toby Conglomerate of Lower Jurassic age is an alluvial-fan deposit of sand, pebbles, and cobbles that built from highlands east of the rift valley. You have the feeling that you are standing inside the fan. The Department of Natural Resources Conservation at UMass-Amherst manages the 755-acre Mount Toby Demonstration Forest for teaching and research.

How to get there. – Drive north from Amherst on Route 116. Proceed about 6 miles to the burial grounds with a white fence on the left (west) side of the highway. Go just past the burial ground and park in the lot near the house set back from the road (453 Long Plain Road). The Roaring Brook trail is adjacent to the parking area on the north side. Lock your vehicle before heading up the trail. Parking is on private land, and the owners maintain the parking area to accommodate hikers. Please pay the requested $1 donation.

Comments. – After walking a short distance to the west up the trail from the parking lot, and just before reaching the railroad track, stop at the small outcrop of dark-colored rock on your left (south). The rock is the Erving hornblende schist of Lower Devonian age. The gray mineral is plagioclase and the dark is hornblende. The schist contains the mineral sillimanite that indicates it formed at about 550 degrees C at about 6 km depth in the Earth's crust. Before metamorphism the original rock was volcanic (Mike Williams, personal communication).

Pick up a small piece of the hornblende schist to carry along with you so that when you get to the Mount Toby Conglomerate you can look for a pebble of it in the conglomerate. None has ever been found here even though the highland source of the conglomerate was just east of the border fault. The pebbles and boulders in the conglomerate are mostly quartz, feldspar, mica, and low to medium-grade schistose metamorphic rocks that were at the surface in the highlands east of the

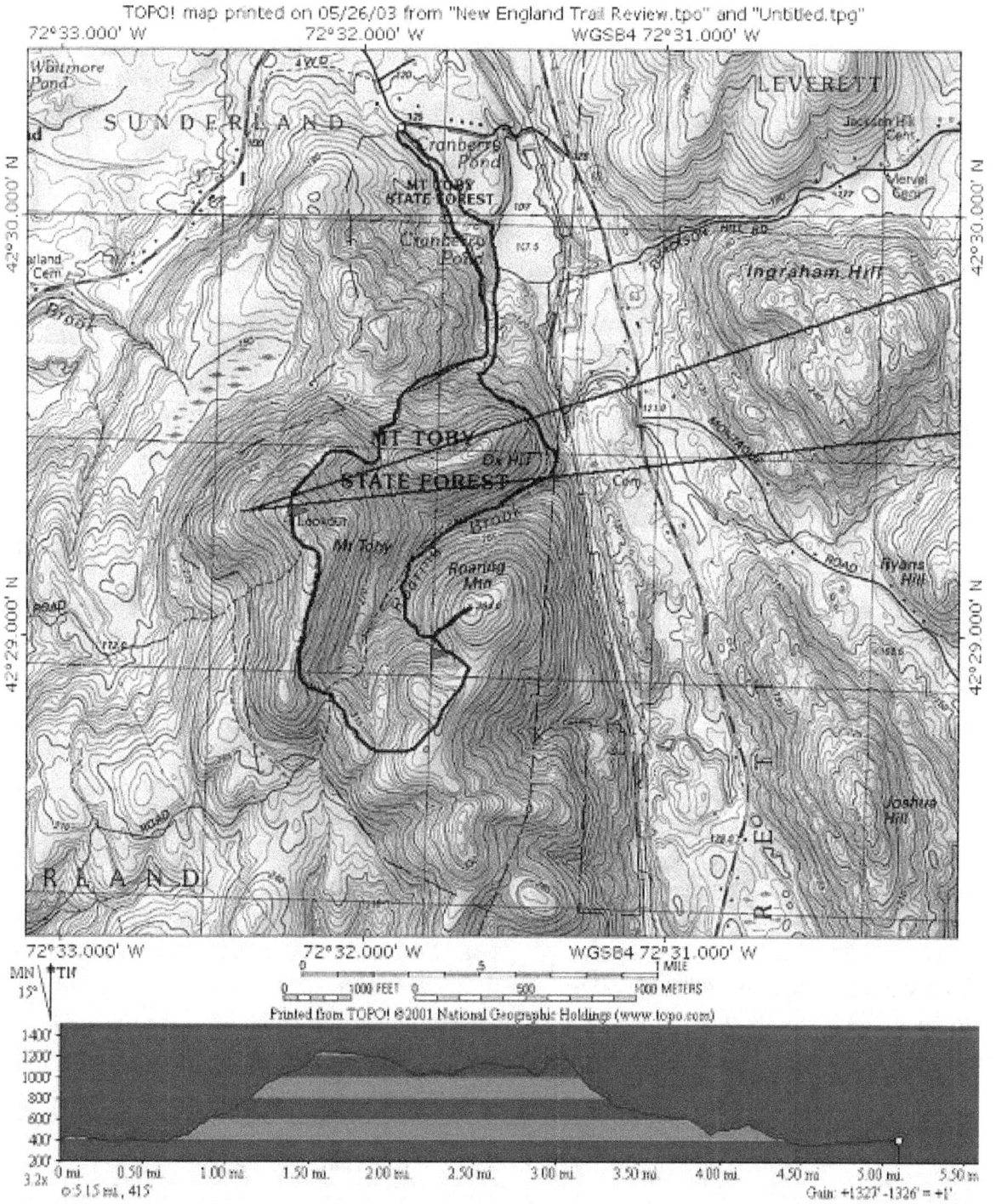

The parking area is on the west side of Route 63 just south of the intersection with Montaque Road, which tends to the southeast. The parking area is equidistant on Route 63 between the two dark lines that show the locations of the elevation profiles. Courtesy of topo.com and the U.S. Geological Survey.

Fire tower on Mount Toby. As a separate trip, you can hike to the fire tower at the 1,269-foot summit of Mount Toby, a four-mile round trip. From Route 2, take Route 63 south and then turn right onto Route 47 south. Proceed for a mile to Reservation Road, which is on your left. The trailhead for the Robert Frost trail (orange blazes) is on your right. Free parking is in a gravel lot, with additional parking available on Reservation Road. Courtesy of Wikipedia.

border fault when the fan was being built. Post-Jurassic erosion has eroded away the highlands, exposing the deep, high-temperature rocks now at the surface east of the fault.

As you leave the hornblende schist, walk to the railroad track, which follows the trace of the border fault.

The two north-south lines subparallel to and west of route 63 on the right side of the image are: 1) on the left, the railroad track that follows the trace of the border fault of the Deerfield basin and 2) a cleared corridor for poles carrying electric power. The Connecticut River and Chard Pond are in the upper left corner. The Mount Toby conglomerate stands in substanially higher relief than the metamorphic rocks east of the fault because it is made of fragments of metamorphic rocks cemented by albite cement. Courtesy of Google Earth.

The fault provides a weakness that allowed erosion to sculpt the valley. The fault dips 40 degrees to the west. The total vertical movement on the fault is about 8 kilometers as evidenced by the thickness of the strata and volcanic rocks in the Deerfield and Hartford basins plus the amount of displacement of igneous intrusions by the continuation of the fault into NH. The crust beneath the basin is thinned by several percent due to the NW-SE extensional pull apart that formed the basin.

As you leave the railroad track, walk up the slope and to the left is Roaring Brook that flows in a valley deepened by glacial erosion. Look up at the "hanging waterfall" that formed because the main valley was deepened more than the tributary. In this photo, the waterfall has substantial discharge, but commonly is virtually dry. Courtesy of http://blogs.umass.edu/bikehara.

In 1973, the Northeast Utilities Company proposed building a nuclear power plant in Montague, but it was canceled in 1980, after spending $29 million on the project. To determine when the fault was last active, three cores were drilled through the fault at Roaring Brook. The zone of shattered rocks along the fault is about three inches wide. Unfractured crystals of the clay-mineral illite had grown across the fault-plane zone of crushed rock. The age of the illite is 144 million years as measured by the K/Ar age-dating method, implying that the fault has not moved since then.

Sam Lovejoy is remembered as leading the opposition to the proposed nuclear power plant. His story is told by Harvey Wasserman in a web site of Ecn.cz.

"In the wee morning hours of February 22, 1974, Sam Lovejoy, a member of one of the local communes near Montague, Massachusetts, slipped onto the Montague Plains and sabotaged the 500-foot weather tower Northeast Utilities had erected to test wind direction at the site. (A company official later explained that the data was needed so that authorities would know which way the radiation would blow from the plant in case of an accident.)

Using a few simple farm tools, Lovejoy left behind him 349 feet of twisted wreckage. He then ran to the nearest road, flagged down a passing patrol car, and got a ride to the Turners Falls station, where he gave Officer Donald Cade the news. In a typed four-page statement, he said:

"As a farmer concerned about the organic and the natural, I find irradiated fruit, vegetables and meat to be inorganic; and I can find no natural balance with a nuclear plant in this or any community. …

I believe that we must act; positive action is the only option left open to us. Communities have the same rights as individuals. We must seize back control of our own community.

The nuclear energy industry and its support elements in government are practicing actively a form of despotism. They have selected the less populated rural countryside to answer the energy needs of the cities. While not denying the urban need for electrical energy (perhaps addiction is more appropriate), why cannot reactors be built near those they are intended to serve? Is it not more efficient? Or are we witnessing a corrupt balance between population and risk? …

Through positive action and a sense of moral outrage, I seek to test my convictions."

Lovejoy went on trial in September, 1974, on charges of malicious destruction of personal property, but the judge dismissed the case on the technicality that the tower was deemed not to be personal property.

At Roaring Brook, be sure to note that the plate-shaped boulders dip upstream. Similar imbrication in the Mount Toby Conglomerate enables measurement of the flow direction. When you arrive at the conglomerate, expect substantial difficulty differentiating bedding planes from the imbrication. Even experienced geologists have this problem.

Measurements of the flow direction based on imbrication of the platy pebbles show that the fan built to just west of north. The fan was a 180-degree cone that faced to the west. The outcrop is a slice through the fan that happens to be close to the trend of the border fault.

Twenty-five years after he toppled the Northeast Utilities weather tower to protest the construction of a nuclear power plant, Sam Lovejoy in 1999 revisits the location of the tower. Courtesy of Masslive.com.

The tower is down. Courtesy of Green Mountain Post Films, Turners Falls, MA.

Walk past the waterfall and up the steep path on the right for a short distance to a prominent flat area in front of a large exposure of conglomerate. You are standing inside a Jurassic alluvial fan.

Each of the size-graded, gravel to sand layers in the conglomerate was deposited by a single storm event. The conglomerate records hundreds of floods that roared down from the mountains located east of the border fault. As each flow decelerated and spread over the fan, the larger stones settled out first, producing a graded bed.

Antidune cross-beds are absent at Mount Toby because this location is near the apex of the alluvial fan, which was adjacent to the border fault. Antidunes formed lower on the fans, as at Chard Pond and West Gill Road.

On a point bar on the Connecticut River at Turners Falls, the imbricated plate-shaped clasts dip to the left, showing that flow was to the right. Courtesy of the Massachusetts Geological Survey

An alluvial-fan conglomerate with particularly well-developed imbrication that dips to the right, indicating flow to the left. Courtesy of William W. Little and slideshare.net.

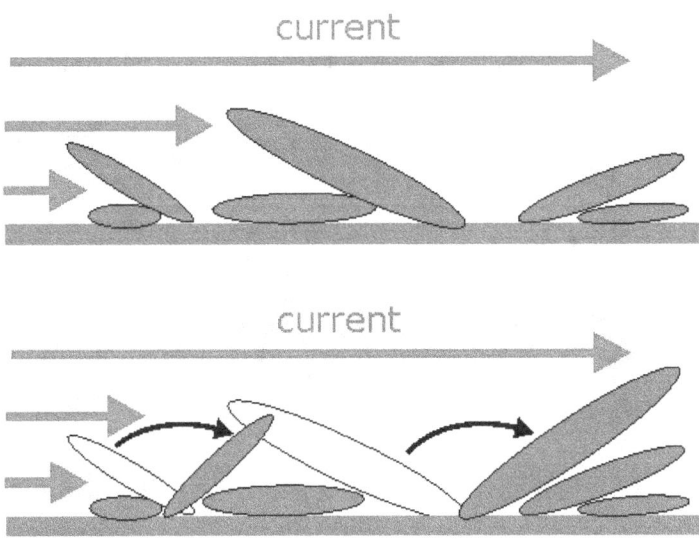

Imbrication forms when disc-shaped clasts with various orientations are flipped over to dip in the up-flow direction. Courtesy of all-geo.org.

A typical view of the conglomerate at Roaring Brook where it is difficult to differentiate bedding from imbrication of the plate-shaped clasts. The bedding is inclined 22 degrees down to the left, as seen in the lower right. Most of the plate-shaped pebbles dip to the left, indicating that the floods moved to the upper right. With a little imagination, the base of a flood unit extents from the base of the largest block in the upper right to the base of a large block in the lower left. Better defined flood units in the Mount Toby Conglomerate are illustrated in the stop at Chard Pond. Courtesy of Paul Karabinos and gigapan.com.

Searches over the years have failed to locate any gravity-driven, debris-flow conglomerate, recognizable by a fabric of pebbles supported by a finer-grained, muddy sand matrix. Debris flows are common on alluvial fans in semi-arid to humid climates, but absent here. This reflects the lack of mud produced during weathering of the crystalline rocks in the source mountains.

Here the bedding and imbrication of the platy clasts are both hard to see. The bedding slopes 10 degrees to the left as seen in the lower left. The imbrication dips to the left so the floods were to the right. Courtesy of Paul Karabinos and gigapan.com.

Deposit of the September, 2013, flash flood on the parking area and road of the alluvial-fan area of Rocky Mountain National Park, Colorado. The imbrication dips to the right and the flood moved to the left. Stones piled up at the "West Alluvial Fan Parking" sign, leaving an empty space behind the sign. Courtesy of GlichChaos and youtube.

14. Deerfield Basalt, Route 2, Gill

What to see. – You can examine almost all of the 55-m-thick Deerfield Basalt, which consists of two early Jurassic lava flows.

How to get there. – If you are on I-91, take exit 27 and proceed east on Route 2 for about 2 miles. Look for the parking area on the right at the Turners Falls dam overlook, located across the street from the basalt outcrop. Be careful crossing the street. It's best to cross at the traffic lights because trucks come downhill from the west at high speed. If you take photographs of the basalt, be sure not to back up into the highway while thinking about the rocks.

To see the basalt pillows in lower flow, walk from the parking area on the sidewalk leading over the bridge to the outcrop just beyond the end of the sidewalk. **Watch out for the traffic.**

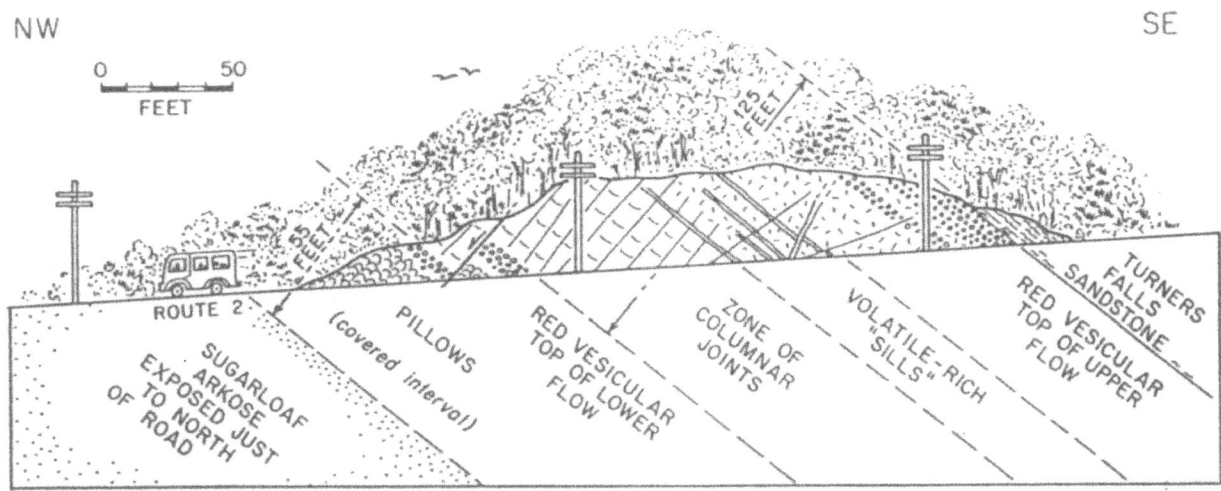

Sketch of the Deerfield Basalt at the Route 2 site (Wise et al., 1992). The pillows at the base of the lower flow are mostly covered by plants. Courtesy of Don Wise.

The contact between the second flow and the overlying redbeds of the Turners Falls Formation is on the right behind the pole. A northeast-trending normal fault is present in the basalt (dark line sloping down to the left corner). The rocks above the fault moved about 15 feet down to the northwest. The orientation and direction of motion on the fault mirror the much larger eastern border fault of the basin.

Comments. – I suggest you start at the base of the basalt and walk up through the two flows. The lower flow is 17-m-thick with a zone of poorly exposed pillow basalt at the base. The upper part of the lower flow has a zone of vesicular bubbles where gases escaped from the surface. The vesicular zone is oxidized to red iron-oxides.

Professor Kurt Hollocher of Union College shows students the pillows and fragments of pillows just above the base of the Deerfield Basalt, which is in the shadow on the lower right. Courtesy of Kurt Hollocher.

The pillow structures formed when the 1700-degree **Fahrenheit** lava entered a 70-degree, shallow lake located in the northeastern corner of the basin. The huge temperature difference causes a lava tongue to cool quickly, forming a thin crust of glass. As the gases push the lava forward, the tongue continues to lengthen, forming a lobe with a pillow-shape in cross section. Eventually the gas pressure within the lobe causes it to burst open to form a new lobe. The result is an interconnecting mass of lobes, each pillow shape in cross-section. In the cold lake water, the glass crusts shatter and fragments are present in the matrix around the pillows. Over millions of years, the glass recrystallizes to a fine-grained crystalline texture.

Basalt pillows in a fragmental matrix are present five meters above the base of the lower flow. The bedding dips to the left at 38 degrees. The stratigraphically up direction is to the upper left. Meter stick for scale. Photo by Jim Dutcher.

Over time, the thin, originally glass crusts on these Deerfield Basalt pillows have altered to dark-colored rock. The stratigraphically-up direction is from lower left to upper right. The photo was taken by Joe Kopera at Highland Park in Greenfield. An automobile key gives the scale. Courtesy of the Massachusetts Geological Survey.

Deerfield Basalt pillows. Courtesy of Planetward.com

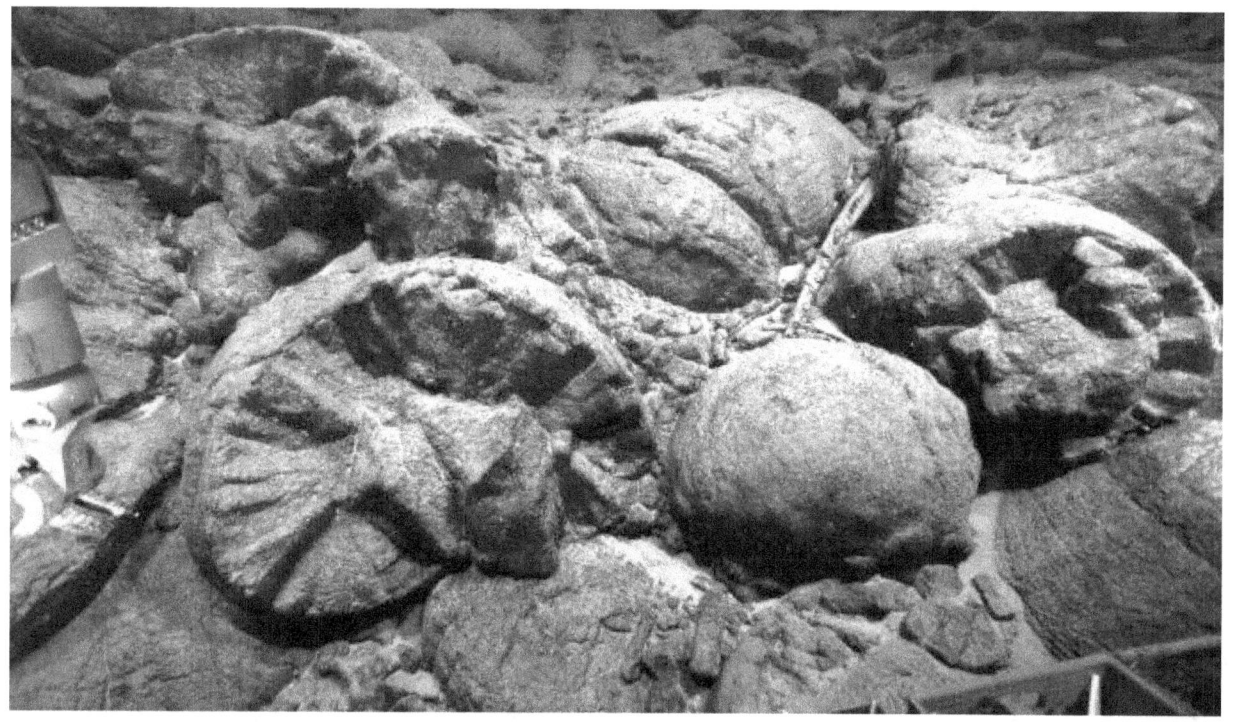

Recently formed pillow basalt on the seafloor near Oahu in Hawaii. Courtesy of the University of Hawaii at Manoa.

The upper flow is 38 meters thick. The lower half is massive basalt, divided into columnar joints that are the result of the liquid flow cooling to solid basalt that occupies less volume. The joints are only in the lower half of the flow, implying they propagated upwards from the base.

The lavas flowed downhill from the western side of the rift valley as shown by regional mapping of basalt-flow directions, using 1) pipe vesicles titled in the flow direction at the base of the flows, and 2) the direction of motion of folds in the then unlithified Fall River muddy beds immediately below the lower lava. At the Route 2 outcrop, flow was to the northeast. A map of the basalt flow directions together with the paleoflow directions of the fluvial redbeds in the Sugarloaf Arkose is in the description of the Route 116 outcrop at Sugarloaf Mountain.

The Deerfield and Holyoke basalts consist of the same two flows, fed by fissure eruptions of the Buttress-Ware dike exposed in Connecticut and Massachusetts. Volcanic cones were rare in the early Jurassic scene. The basalts were later separated by erosion near Amherst into the Deerfield and Hartford basins.

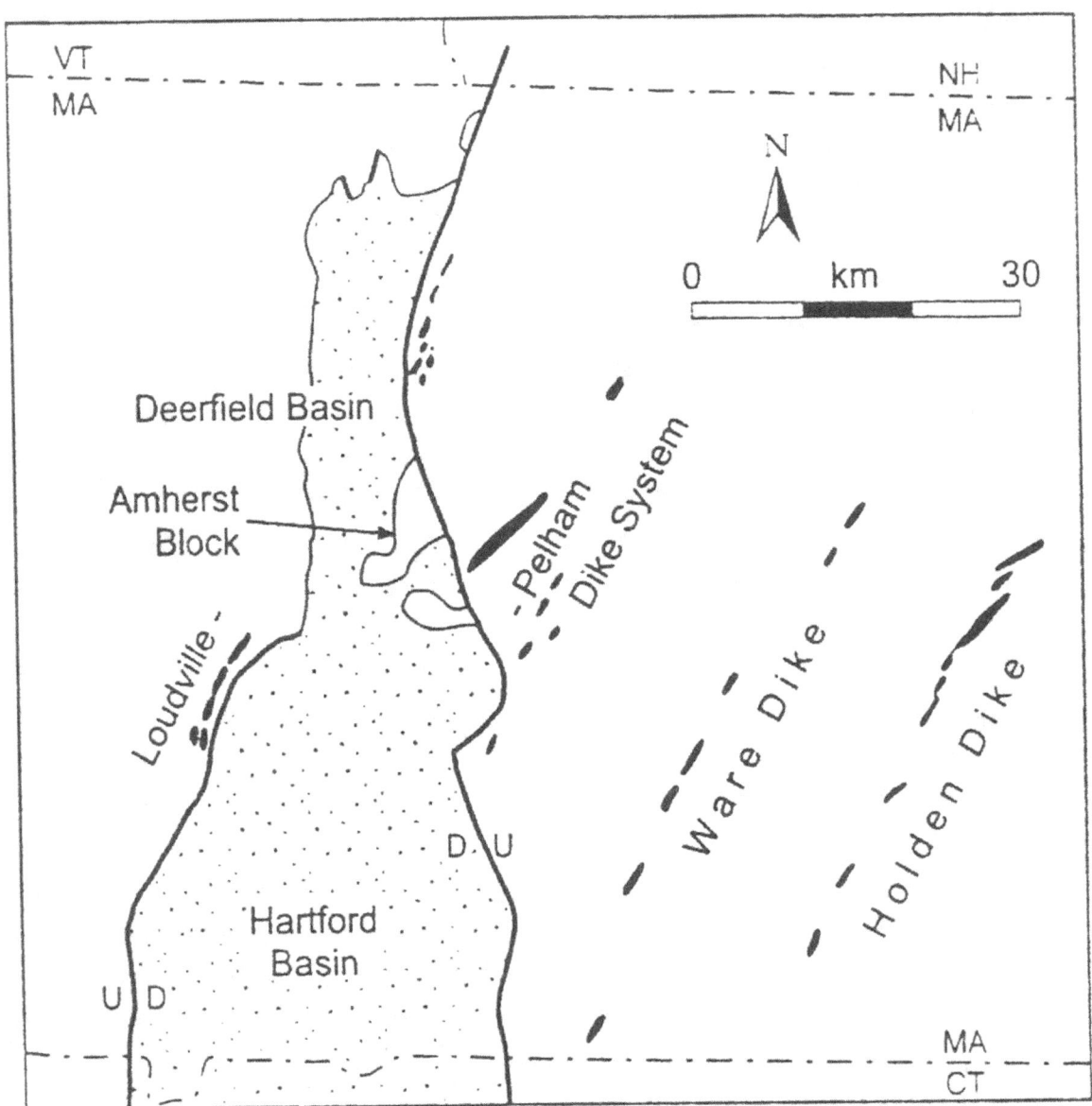

The Deerfield basalt lavas came from fissure eruptions along the Ware-Buttress dike that crops out east of the Deerfield Basin and west of the Hartford Basin. The lavas covered an extensive area in Massachusetts and Connecticut, including the terrain west and east of the Deerfield Basin. The basin was similar to modern rift valleys with a fault on one side. The border fault on the up-thrown (east) side resulted in a tilted block that sloped eastward away from the rift. As the lavas flowed from the Ware fissure, the fault-bounded highlands east of the rift blocked lava from entering the basin. The result was that the west side of the basin affected a "capture" of the lavas, causing them to flow east down into the basin, pinching out on the west-facing slope of alluvial-fans. From Hubert and Dutcher, 1999.

Fountains playing along a fissure eruption in Iceland. Courtesy of thesundaytimes.co,uk.

Pipe vesicles at the base of a lava flow were tilted in the direction of flow by the forward motion of the lava in South Africa. Courtesy of researchgate.net.

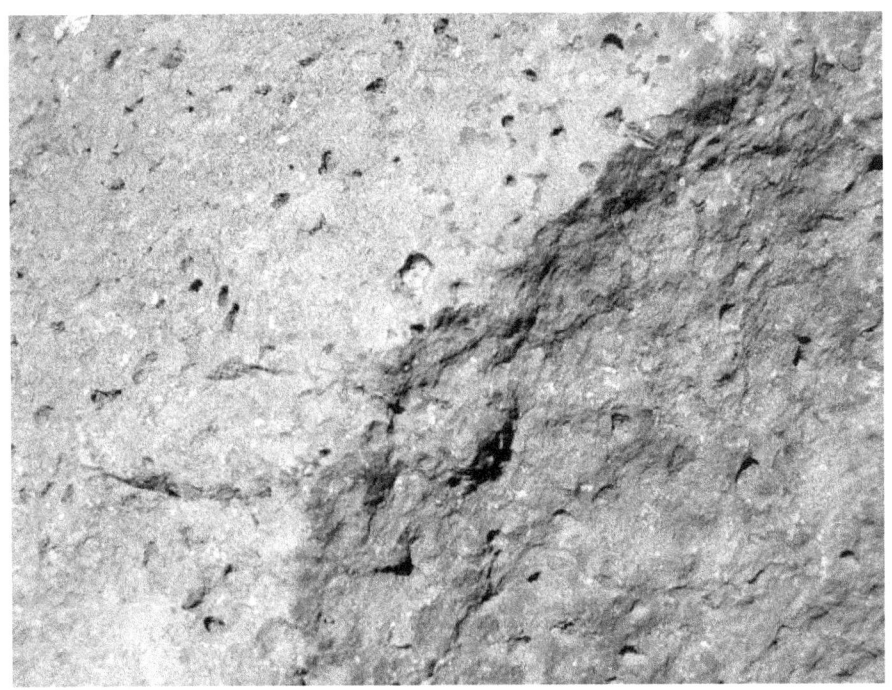

Escape of gases from the surface of the flow left vesicles in the basalt. Courtesy of Kurt Hollocher.

Gases pour from the Holuhraun fissure eruption at Iceland's Bardarbunga volcano in September, 2014. The gases are mostly water vapor, carbon dioxide, sulphur, and sulphur dioxide. Courtesy of national radio programs of abc.net.

The contact between the second flow and the layered sandstone of the Turners Falls Formation.

In the columnar jointing of the second flow, the lines from lower right to upper left are fracture steps formed during pauses in cooling as the columnar joint propagated from lower left to upper right. Photos courtesy of Kurt Hollocher.

15. Sugarloaf Arkose and Hydrothermal Hot Spot, Greenfield

What to see. – The cliff exposes the upper 84 feet of the fluvial Sugarloaf Arkose and the unconformity between it and the 10-foot-thick Fall River beds (marsh, playa, lake). Your first impressions are that the Sugarloaf Arkose is not the usual brownish-red and the normally brownish-red mudstones are nearly black. This is a "hot spot" where the sandstones were bleached by reaction with hot waters rising from deep in the basin about 185 million years ago.

On the Massachusetts state geologic map, the Triassic/Jurassic boundary at 206 million years is arbitrarily drawn 100 m below the base of the Fall River beds, which elsewhere contain early Jurassic spores and pollen. The age of the Deerfield Basalt is 201 million years.

How to get there. – The outcrop is behind the Super Stop and Shop at 416 Federal Street, Greenfield. Park in the store parking lot and walk to the rock wall excavated to make room for the supermarket.

Do not climb up the rock face or the store manager will tell you to leave and never come back. Be careful to **watch for trucks** going to and from the loading dock behind the store.

Comments. – **Sugarloaf Arkose**

The Sugarloaf Arkose was deposited by a braided river with a gravel-sand bedload. The 22 channels mostly are 0.5 to 1 m deep. The channel fills are 80% horizontally-laminated coarse sandstone and pebbly sandstone deposited during shallow, rapid flows. The 20% cross-bedded sandstones formed from trains of subaqueous dunes that filled low areas in the channels. You can

High up on the rock wall is the slight angular unconformity with three feet of erosional relief between the Sugarloaf Arkose and the overlying 10-foot-thick Fall River beds, which are darker and finer-grained than the arkose. The unconformity records the change from an open, fluvial basin with a river flowing out of it to a topographically closed basin with playas and lakes.

The outcrop has recently been landscaped with white pines placed on small shelves with almost no soil. Perhaps a bit more pleasing to the eye, but a loss to geology. A meter stick is present in the lower left. Photo by Jim Dutcher.

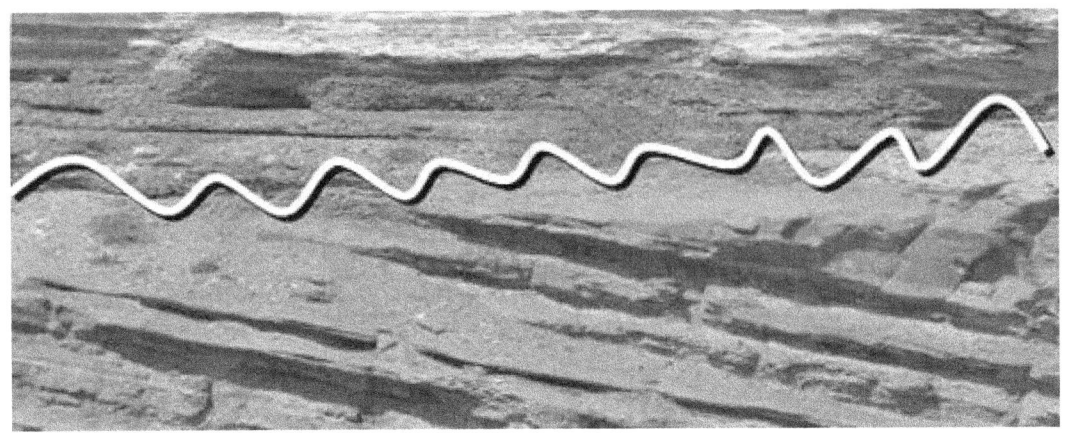

The sequence in an angular unconformity is 1) deposition of the lower sedimentary sequence in horizontal orientation, 2) tilting of the sequence, 3) erosion, and 4) deposition of the upper sequence. Courtesy of Slideplayer.US.

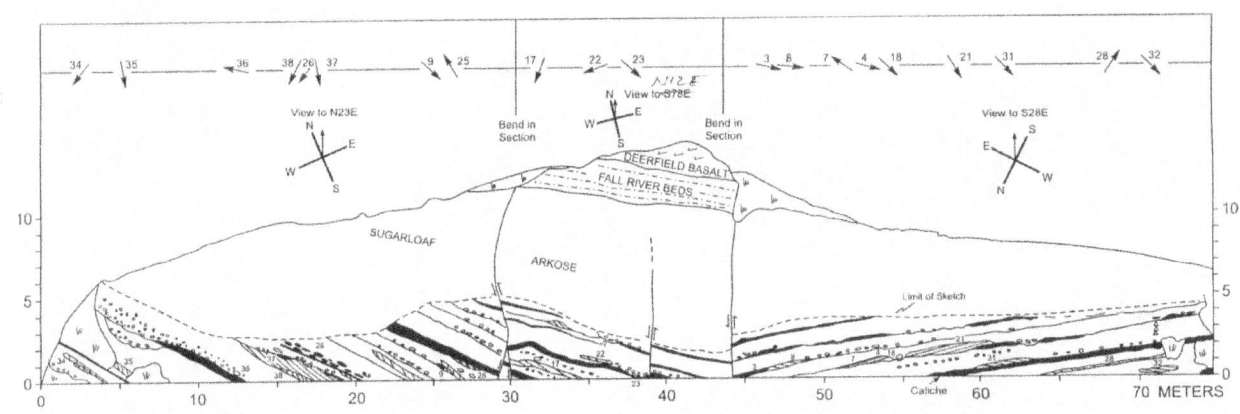

Sketch of the outcrop made before the extensive landscaping, showing stratigraphy, rock types, and paleocurrents. Channel sandstones are white; floodplain mudstones are black. The outcrop comprises three continuous sections, separated by two bends in the rock wall.

Each paleocurrent arrow along the line at the top of the sketch represents one paleocurrent reading, mostly cross-beds. The arrow shows river-flow direction relative to the direction the viewer is looking. The number with the arrow refers to the number assigned to the reading. The river flowed mostly to the southwest. In the left and middle sections, paleoflow in the river channels was approximately towards the viewer so that cross sections of channels are clearly visible. In the right-hand section, paleoflow more closely parallels the rock face. From Hubert and Dutcher, 1999 (modified from Lounsbury, 1990).

easily see cross-beds on the right-hand end of the outcrop. Cross-beds thicker than 40 cm are inferred to be channel bars.

The channels are bordered by floodplains where 4-60 cm-thick layers of sandy mudstone were deposited during floods. The ratio of sandstone to mudstone is 85:15, but maybe half of mudstone was removed by channel scour. Semiaridy is implied by very small soil nodules of calcite (calcium carbonate) in the mudstone. A few beds have mm-scale root casts.

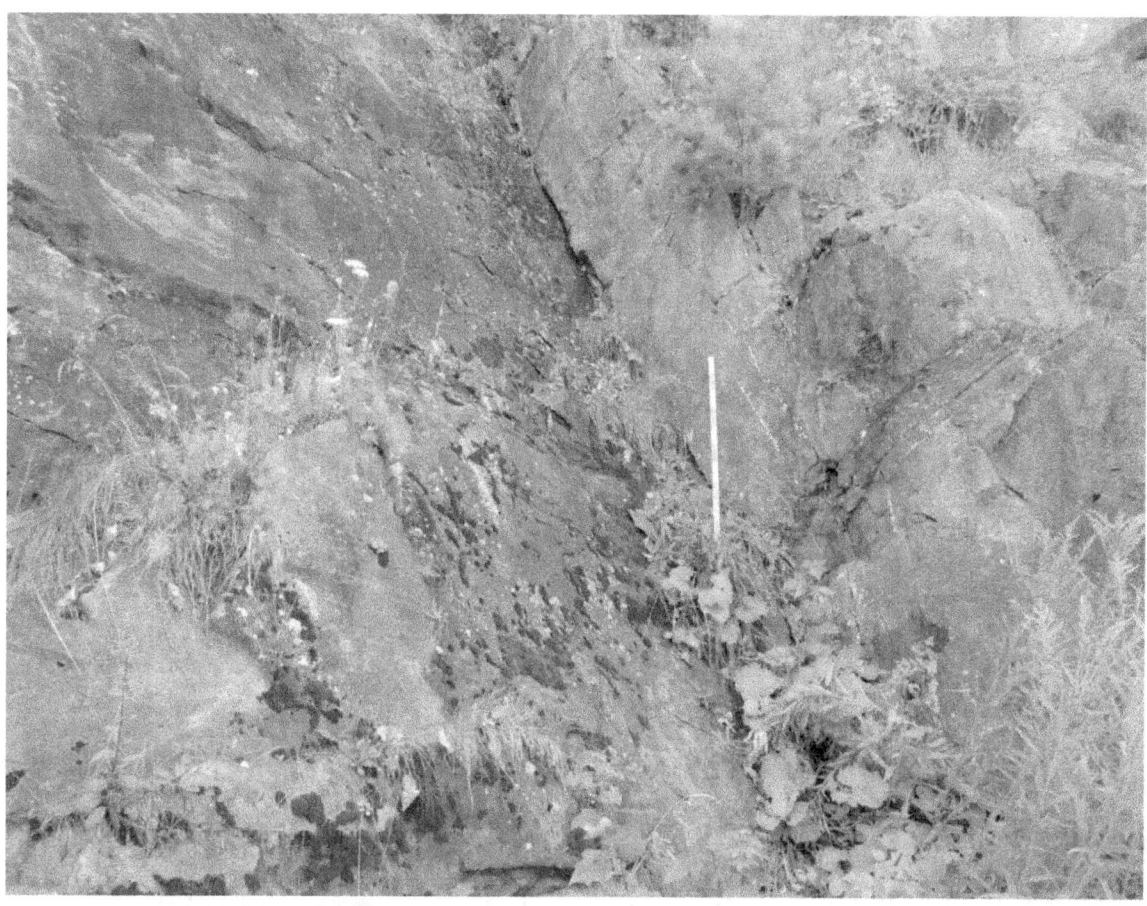

The surface of a steeply-dipping, normal fault fills the left half of the photo. The base of the meter stick (10-cm intervals) rests between the two fault blocks. The rocks above the fault moved down to the right, displacing a dark overbank mudstone (mid-stick) about a meter (bottom of stick). A reverse fault would have the opposite sense of motion. The fault trends to the northeast and the motion is down to the northwest, similar to the orientation and sense of motion of the master border fault on the east side of the basin. A light-colored patch of limonite and small crystals of quartz (upper left) was precipitated on the fault plane. The fault extends through the Deerfield basalt. Photo by Jim Dutcher.

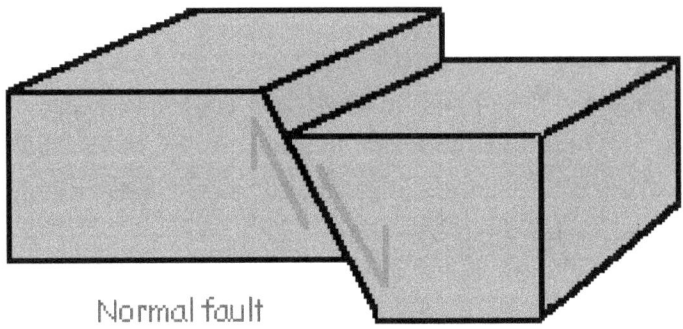

In a normal fault, the rocks above the fault move down relative to the rocks below the fault. Courtesy of mbmg.mtech.edu.

In this stack of three cross-bed sets deposited in a river channel, the lowest cross-bed set dips to the right and the overlying two dip to the left. The scale is in centimeters and inches. Photo by Jim Dutcher.

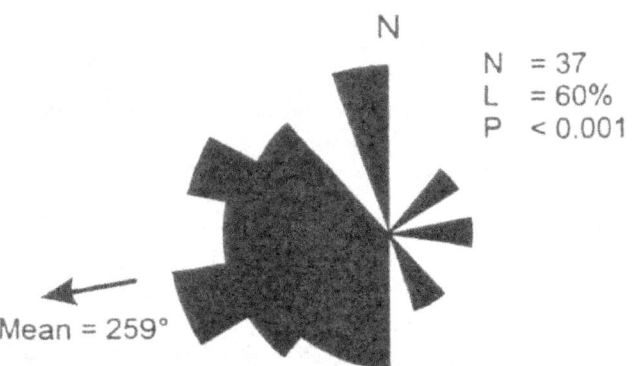

Plot of the 37 paleoflow readings, showing that the river on average flowed to the southwest towards an azimuth of 259 degrees. The probability value (P) says there is less than one chance in a thousand that this strong a preferred origin could arise by chance alone. If all the readings went in the same direction, the consistency ratio (L) would be 100 percent.

The glacial striations from lower left to upper right were protected from modern weathering until the soil was removed during construction of the store. Stones in the ice made the scratches when the ice moved slowly forward by plastic deformation. The rock surface is rounded, polished, and grooved. Photo by Jim Dutcher.

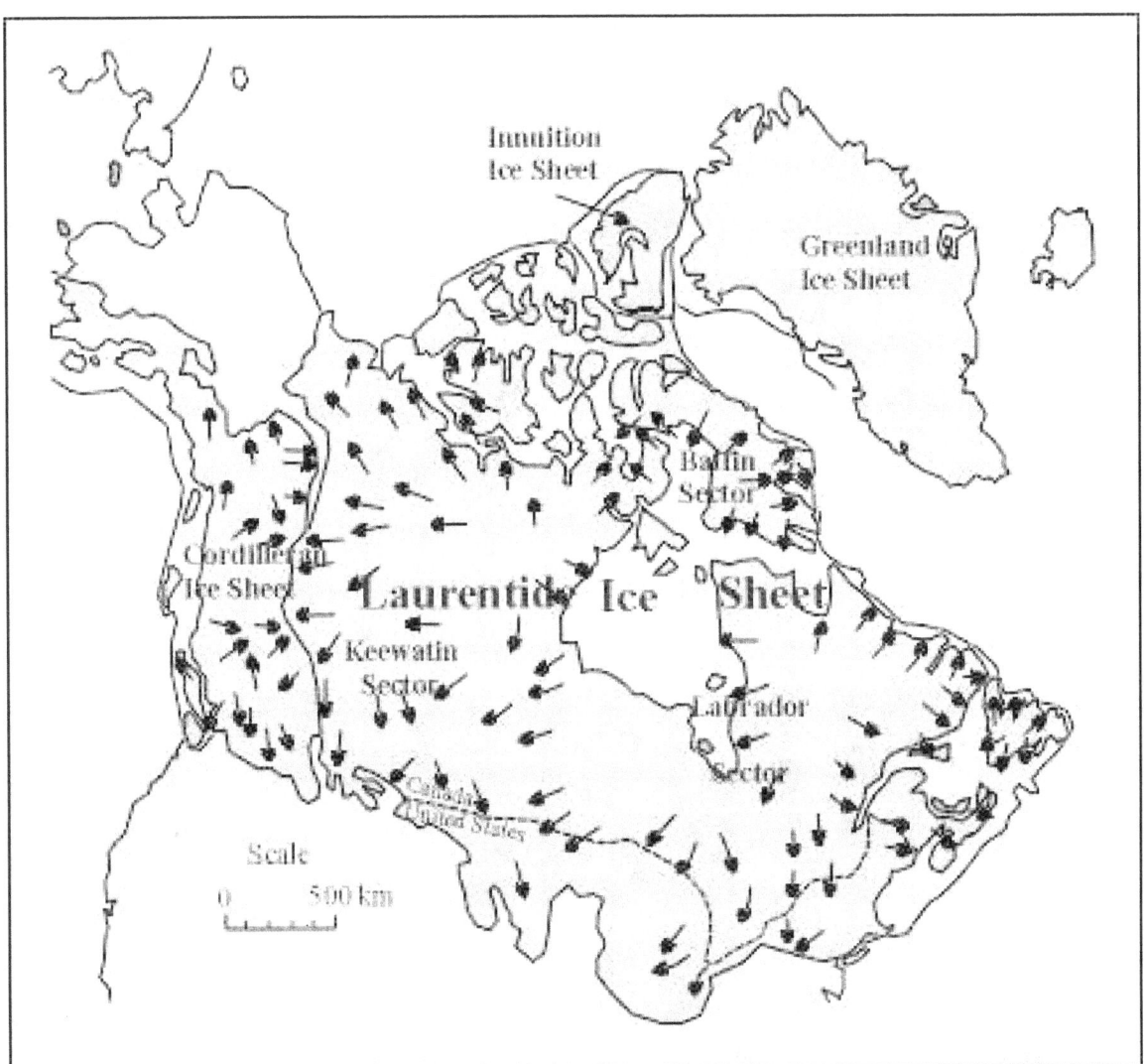

The glacial striations were made by stones transported in the Laurentide ice sheet 14,000 years ago as the ice expanded southward. In western Massachusetts, the dominant direction of ice advance was northwest to southeast. Fainter striations were locally superimposed over the early ones by a later ice readvance from north to south. The ice buried the Holyoke Range and Sugarloaf Mountain. Courtesy of block6sciencetour.wordpress.com.

What were the rivers like that deposited the Sugarloaf Arkose on the rift valley floor, away from the fan built from highlands west of the rift? Unfortunately, the answer is complex because rivers defy simple generalizations. At any one time, a river along its path has various channel morphologies due to interactions among valley slope, discharge, bedload, and other factors

(Schumm, 2005). For the Sugarloaf Arkose, the problem is compounded by the 16 million years of accumulation when river channel morphology varied due to changes in controlling factors, plus the variations associated with the 20-km down-river extend of the outcrops.

Despite these complications, the rivers mostly were gravel/sand bedload-overbank systems in a climate where the dry season dominated over the wet. Most of the outcrops indicate that the rivers were braided, for example here, at Sugarloaf Mountain, and Country Club Road. Tightly meandering river patterns were rare or absent as evidenced by the lack of several criteria, namely 1) lateral accretion surfaces (sloping layers of sand) formed on point bars at meander bends, 2) fine-grained sandstones and siltstones attributable to crevasse splays where a river breaks though a levee to deposit muddy sediment on the adjacent floodplain, and 3) stacked, grain-size fining up channels caused by the passage of meander bends over a specific location.

Fall River beds

The Fall River beds are brownish-red and gray mudstones and fine-grained sandstone deposited in a shallow lake and marsh. The lake deepened northward towards Route 2 and towards the border fault on the east side of the basin.

Deerfield basalt

At the right-hand end of the outcrop is a former path now recognizable as a swale in the grass that leads up slope to the poorly exposed Deerfield Basalt. The lower of the two flows has a 20-m-thick zone of elongate pillow structures, some jumbled, others twisted.

Before the landscaping, folds were exposed in the Fall River beds immediately below the base of the basalt. The unconsolidated muddy sands were deformed into folds as the lava flowed downslope into a shallow lake. The heights of the folds are 10 to 60 cm and the distances between fold crests are about 1 to 2 m. The axes of the folds are perpendicular to the flow direction and folds are tilted in the direction of flow to the northeast. The regional pattern of lava flow is down slope into a low area just west of the border fault.

Hydrothermal Alteration

The Deerfield basin experienced an increased rate of heat flow in the early Jurassic associated with a more rpid rate of crustal extension and thinning of the Earth's crust. The fill of the basin is about 4 km with 2 to 3 km eroded off the top. A typical increase in temperature with depth for a rift basin is about 40 degrees centigrade per km. Six kilometers produces a temperature at the bottom of the basin of 240 degrees centigrade. The high temperature generated a reservoir of hot brine with substantial dissolved sodium in the sands deep in the basin. The sodium mostly came from partial dissolution of sodium-rich feldspar grains in the arkose. About 185 million years ago, rising plumes of hot water moved up though the basin strata, producing the hydrothermal alteration at this outcrop.

At the far-left end of the outcrop, hidden in the plants, are pale-colored sandstones with very tiny dark spots. These are clumps of chlorite crystals of a distinctive type that only forms at temperatures between 150 to 200 degrees centigrade.

Basalt pillow at Super Stop & Shop, Greenfield. The stratigraphically-up direction is to the upper right. The meter stick is divided into centimeters. Photo by Jim Dutcher.

16. Geo Path and Rock Park, Greenfield Community College

What to see. – With a variety of igneous, metamorphic and sedimentary rocks, this site is a refresher course in their identification and origin. Of special interest are armored mudballs from the Sugarloaf Arkose and Turners Falls Formation.

How to get there. – Exit I-91 at Route 2 in Greenfield and proceed west on Route 2. Almost immediately take a right to the north on Collrain road. At the traffic circle go west on College Drive to the college. The start of the Geo Path is adjacent to the parking area at the Sloan Theater and leads to the rock garden in the northwest corner of the large, sprawling main building.

Comments. – The Greenfield Community College Rock Park was begun in the early 1980s by Professor Emeritus Dick Little of GCC to preserve armored mud balls that he discovered in the foundation of a former bridge across the Connecticut River at Unity Park, Turners Falls. The town highway department separated the blocks with the armored mudballs and they were transported to GCC. Large samples of other rock types were added and the Rock Park officially opened in 1985.

In 2016, about 20 samples were moved to the Geo Path because maple shade trees in the Rock Park were dripping sap on many of the rocks, obscuring the geologic features.

The boulders in the Geo Path and the Rock Park comprise a fine display of sedimentary, igneous and metamorphic rocks. The sedimentary rocks include shallow-marine fossils present in lower Paleozoic (Ordovician to Devonian) limestones from the Heidelberg Mountains of Eastern New York. There are pencil-sized coral colonies, larger "horn" corals, small colonies of branching bryozoans, clam-like brachiopod shells, and crinoid stems plus the fossils of other animals. At this time, eastern North America was in the tropics south of the equator.

At the start of the Geo Path, the large white boulder in the lower right is quartzite whose surface is marked with crescentic fractures caused by collisions with other rocks during stream transport. The boulder was carried in glacial ice from southern Vermont. The first five boulders in the upper left are pebbly sandstones with armored mudballs originally from the Sugarloaf Arkose and Turners Falls Formation.

On the continuation of the Geo Path, the boulders from lower right to upper left are: granite porphyry; gneiss with a granite vein; mica schist with a coarse-grained granite intrusion; and white marble. Photos courtesy of Dick Little.

Left. An armored mudball in a pebbly sandstone where the fast-moving current could roll the pebbles. The bedding in the pebbly sandstone is from lower left to upper right. Dick Little first discovered armored mudballs, including this one, in abutments made of Turners Falls sandstone from the former bridge from Montague to Gill (Dick Little, 1982). The Greenfield highway department dismantled the bridge foundation, which allowed the mudballs to be moved to the GCC campus. The bridge was built in 1878 and destroyed in the flood of 1938. In his web site http://earthview.rocks/mudballs.html, Dick explains, "The valley armored mudballs formed when large pieces of hard, dry mud fell into a steam. As they tumbled in the current, they became round as well as soft and sticky enough on the outside to have streambed pebbles imbed into the soft exterior, forming the armor. To be preserved, the newly created armored mudballs must be buried quickly in the stream's gravel sediment." *Right*. Dick Little points to two armored mudballs. Courtesy of Dick Little.

One of many armored mudballs that formed during a flood event in a stream in china. From Gerhard H. Bachmann (1942).

Granite porphyry is a term for an igneous rock consisting of large-size crystals of feldspar and quartz dispersed in a fine-grained matrix. The large white crystals are feldspar that formed early in the crystallization of the liquid magma, where they were able to grow to large sizes.

Blocks of Deerfield Basalt where the tall specimen is a six-sided column from columnar basalt. Dick Little stands between two basalt columns. Photos courtesy of Dick Little.

The metamorphic rocks illustrate mica schist, gneiss, quartzite, and marble formed by the alteration of pre-existing rocks by heat, pressure, and chemically-active fluids during the continental collisions that occurred as the supercontinent of Pangea was assembled in the late Paleozoic. Pebbles of many of the types on display are present in the conglomerates of the Deerfield basin.

The intrusive deep-origin igneous rocks that cooled from magma include granite, pegmatite, and vein quartz, all common pebbles in the Deerfield basin. **The** extrusive, lava-flow igneous rocks are mostly from the Cheapside Quarry in the Deerfield Basalt located in east Deerfield.

References Cited

Alexander, R.M., 1976, Estimates of speeds of dinosaurs: Nature, v. 261, p. 129-130.

Bachmann, G. H, 2014, Armored mud balls as a result of ephemeral fluvial flood in a humid climate: modern example from Guizhou Province, South China: Journal of Palaeogeography, v. 3, p. 410-418.

Brigham-Grette, Julie, and Wise, D. W., 1988, Glacial and deglacial landforms of the Amherst area: *In* Julie Brigham-Grette, ed., *AMQUA Field Trip Guidebook*: Contribution No. 63, Department of Geology and Geography, Univ. of Massachusetts at Amherst, p. 209-244.

Coombs, W. P, 1980, Swimming ability of carnivorous dinosaurs: Science, v. 207, p. 1198-1200.

Curren, K., 1999, Landscape evolution revealed by archeological excavations at Peskomskut: Master's Thesis, Department of Anthropology, University of Massachusetts at Amherst.

Farlow, J. O. and Galton, P. M., 2003, Dinosaur trackways of Dinosaur State Park, Rocky Hill, Connecticut: *In* P. M. Letourneau and P. E. Olson, ed., *The Great Rift Valleys of Pangea in Eastern North America*: Columbia Univ. Press. v. 2, p. 248-263.

Galton, Peter and Farlow, J. O., 2003, Dinosaur State Park, Connecticut, USA: history, footprints, trackways, exhibits: Zubia, number 21, p. 129-174.

Getty, P. R., Judge, A. I., Csonka, Jayme, and Bush, Andrew, 2012, Were Early Jurassic dinosaurs gregarious?: Reexamining the evidence from dinosaur footprint reservation in Holyoke, Massachusetts: *In* Margaret Thomas, ed. *Guidebook for Fieldtrips in Connecticut and Massachusetts*: State Geological and Natural History Survey of Connecticut, 47th Annual Meeting of the Northeastern Section of the Geological Society of America, p. A1-A18.

Harms, J. C., Southard, J. B., Spearing, D. R., Walker, R. G., 1975, *Depositional Environments Interpreted from Primary Sedimentary Structures and Sequences*: SEPM, Dallas, Texas, 161 pp.

Harms, J.C., Southard J.B., and Walker, R.G., 1982. Structures and sequences in clastic rocks. SEPM Short Course 9.

Hubert, J. F. and Dutcher, J. A., 2010, *Scoyenia* escape burrows in fluvial pebbly sand: upper Triassic Sugarloaf Arkose, Deerfield rift basin, Massachusetts, USA: Ichnos, v. 17, p. 17-20.

Hubert, J. F., Dutcher, J. A., and Walsh, M. P. 2008, Tectono-sedimentary history of the Late Triassic New Haven Arkose, Deerfield rift basin, Massachusetts: Northeastern Geology and Environmental Sciences, v. 30, p. 354-374.

Hubert. J. F. and Dutcher, J. A., 2005, Synsedimentary sand pillows on a lacustrine delta slope and sheet-flood deposition of alluvial-fan gravels, early Jurassic, Deerfield basin, Massachusetts: Northeastern Geology and Environmental Sciences, v. 27, p. 18-36.

Hubert, J. F., Taylor, J. M., Ravenhurst, C., Reynolds, R., and Panish, P. T., 2001, Burial and hydrothermal diagenesis of the sandstones in the early Mesozoic Deerfield rift basin, Massachusetts: Northeastern Geology and Environmental Sciences, v. 23, p. 109-126.

Hubert, J. F. and Dutcher, J. A., 1999, Sedimentation, volcanism, stratigraphy, and tectonism at the Triassic-Jurassic boundary in the Deerfield basin, Massachusetts: Northeastern Geology and Environmental Sciences, v. 21, p. 188-201.

Hubert, J. F., Feshbach-Meriney, P. E., and Smith, M. E., 1992, The Triassic-Jurassic Hartford rift basin: evolution, sandstone diagenesis, and hydrocarbon history: American Association of Petroleum Geologists Bulletin, v. 76, p. 1710-1734.

Hubert, J. F., and Reed, A. A., 1978, Redbed diagenesis in the East Berlin Formation, Newark Group, Connecticut Valley: Journal of Sedimentary Petrology, v. 48, p. 175-184.

Jenkins, Paul, 1982, *The Conservative Rebel: A Social History of Greenfield, Massachusetts*: Town of Greenfield, Massachusetts, 287 pp.

Kent, D. V. and Tauxe, L., 2005, Corrected Late Triassic latitudes for continents adjacent to North America: Science, v. 31, p. 240-244.

Lepore, Taormina, 2006, New theropod and ornithischian footprints at the Dinosaur Footprint State Reservation (early Jurassic, Portland Formation), Holyoke, Massachusetts, U. S. A:

undergraduate honors project, Department. of Biology, University of Massachusetts at Amherst, MA, 48 pp.

Little, R. D., 1982, Lithified armored mud balls of the Lower Jurassic Turners Falls Sandstone, north-central Massachusetts: Journal of Geology. v. 90, p. 203-207.

Lounsbury, R. 1990, Hydrothermally altered Sugarloaf Arkose, Greenfield, Massachusetts: Undergraduate research report, Department of Geosciences, University of Massachusetts at Amherst, 11 pp.

Lull, R. S., 1953, *Triassic Life of the Connecticut Valley*: Connecticut Geological and Natural History Survey, 336 pp.

McDonald, N. G., 2010, *Window into the Jurassic World*: Friends of the Dinosaur Park and Arboretum, Rocky Hill, Connecticut, 106 pp.

McDonald, N. G., and LeTourneau, P. M., 1990, Revised paleogeographic model for Early Jurassic deposits, Connecticut valley: regional easterly paleoslopes and internal drainage in an asymmetrical extensional basin: Abstracts vol., Northeastern Section Geological Soc. of America meeting, p. 54.

Olsen, P.E., 2010, Fossil great lakes of the Newark Supergroup – 30 years later: *in*: A. I. Benimoff,, ed., *Field Trip Guidebook*: New York State Geological Association, 83nd Annual Meeting, College of Staten Island, p. 101–162.

Olsen, P. E., McDonald, N. G., Huber, Phil., and Cornet, Bruce. 1992, Stratigraphy and paleoecology of the Deerfield rift basin (Triassic-Jurassic, Newark Supergroup), Massachusetts: *in* P. Robinson and J. B. Brady, eds., *Field Trip Guidebook of the 84th Annual New England Intercollegiate Geologic Conference*: Department of Geosciences, University of Massachusetts at Amherst, Massachusetts, v. 2, p. 488-535.

Ostrom, John (1972). "Were some dinosaurs gregarious?" :Palaeogeography, Palaeoclimatology, Palaeoecology, v. 11, p. 287–301.

Philpotts, A. R, 2012, The Holyoke Basalt, its source and differentiation in a thick flood-basalt flow: Geological Society of America Northeast Section meeting, Hartford, CT, Abstracts Vol., p. 56.

Robinson, P. L., 1973, Paleoclimatology and continental drift: *in* D. H. Tarling and S. K. Runcorn., eds., *Implications of Continental Drift to the Earth Sciences,* vol. 1: Academic Press, New York, p. 451-476.

Schumm, S. A., 2005*, River variability and complexity*: Cambridge University Press, Cambridge, U. K., 220 pp.

Robinson, P. L., 1973, Paleoclimatology and continental drift: *in* D. H. Tarling and S. K. Runcorn., eds., *Implications of Continental Drift to the Earth Sciences,* vol. 1: Academic Press, New York, p. 451-476.

Schumm, S. A., 2005*, River variability and complexity*: Cambridge University Press, Cambridge, U. K., 220 pp.

Van Houten, F. B.., 1977, Triassic-Jurassic deposits of Morocco and eastern North America: comparison: American Association of Petroleum Geologists Bulletin, v. 61. p.79-99.

Wise, D. U., Mabee, S. B., and Condit, C. D., 2015, *Field Trip Guidebook of the 66th Highway Geology Symposium in Western Massachusetts:* Massachusetts Geological Survey, Amherst Massachusetts, 85 pp.

Wise, D. U. and Hubert J. F., 2003, Evolving fault types, stress fields, and tectonics of the early Mesozoic Deerfield basin, Massachusetts: *in* J. B. Brady and T. C. John., eds*. Guidebook for Field Trips, New England Intercollegiate Geologic Conference,* Amherst and Northampton, Massachusetts, p. C1-30.

Wise, D.U., Hubert, J. F., and Belt, E. C., 1992, Mohawk trail cross section of the Mesozoic Deerfield basin: structure, stratigraphy, and sedimentology: *in*, Peter Robinson and J.B, Brady, eds*., Field Trip Guidebook of the 84th Annual New England Intercollegiate Geological Conference:* Amherst, Massachusetts., v. 1, p. 170-198.

Taylor, J. M., 1991, Diagenesis of sandstones in the early Mesozoic Deerfield Basin, Massachusetts: Master's Thesis, Department of Geosciences, University of Massachusetts at Amherst, 225 pp.

Zen, E-AN, 1983, Bedrock geologic map of Massachusetts: U.S. Geological Survey, Washington, D. C.

Web Credits

Introduction

Triassic world. http://www.glosgeotrust.org.uk/downloads/pangaea.jpg

Earth's tectonic plates. https://www.e-education.psu.edu/earth520/content/l2_p14.html

Cross section of the Earth. http://www.age-of-the sage.org/tectonicplates/boundaries_boundary_types.html

Map of western Massachusetts. http://www.aaccessmaps.com/show/map/us/ma/mass_w

1. Summit of Mount Tom

Map of Mount Tom. http://www.mappery.com/Mt-Tom-State-Reservation-trail-map

Directions to Mount Tom. http://www.mass.gov/eea/agencies/dcr/massparks/region-west/mt-tom-park-directions.html

Migrating hawks and falcons. http://berkshirehiking.com/hikes/mt_tom_ma.html

Mount Tom. https://www.yelp.com/biz/mount-tom-state-reservation-holyoke

Mount Tom Hotel. http://www.playle.com/listing.php?i=DOLLARSTORE154026

Oxbow marina. https://dmampo.org/aerial-of-oxbow-and-oxbow-marina-on-connecticut-rivernear-holyoke-ridge-and-mount-tom-northampton-ma/

Hang gliding from Mount Holyoke. https://www.mass.gov/locations/mount-holyoke-range-state-park

2. Dinosaur Footprints Reservation, Holyoke

Discovery of dinosaur tracks. https://jurassicroadshow.com/downloads/geologic-walking-tour-of-turners-falls/

Owen's Lake, a flooded playa. http://mavensphotoblog.com/2012/04/07/the-owens-lake-dust-control-project-the-ultimate-human-managed-landscape/

Map of the tracks, http://tumblehomelearning.com/were-dinosaurs-gregarious-or-antisocial-dinosaur-tracks-provide-clues/

Calculation of dinosaur speeds from tracks. mnzoocdn.mnzoo.org/wp-content/uploads/2016/05/D16-Acti-Sheet-Speed.pdf   OMIT

Calculation of dinosaur speeds from tracks. www.amnh.org/content/download/49379/751532/file/dinoactivity_speed.pdf

Dimension-less stride. http://userhome.brooklyn.cuny.edu/grocha/lab/behavior.html

Dinosaur footprints. http. http//explorewmass.blogspot.com/2008/05/visit-to-holyokes-dinosaur-footprints.html-

Guidebook to the Hartford basin. https://www.ldeo.columbia.edu/~polsen/nbcp/LeTourneau+15_facies-fossils.pdf

Primitive cycads. https://www.smith.edu/garden/events-exhibits/exhibits/.../evolution-mural-panels

3. Playa and lacustrine strata, Holyoke Community College

Spinning top sweeps out a cone. https://www-spof.gsfc.nasa.gov/stargaze/Sprecess.htm

Lake cycles in East Berlin Formation. http://www.reasons.org/articles/milankovitch-cycle-design
Earth's elliptical cycle.
http://www.indiana.edu/~geol105/images/gaia_chapter_4/milankovitch.htm

4. Dinosaur Park and Arboretum, Rocky Hill

Eubrontes. https://www.ldeo.columbia.edu/~polsen/nbcp/eubrontes.html

Dinosaur footprints. alltrails.com/trail/us/massachusetts/dinosaur-footprints

Photos of Dinosaur State Park, Rocky Hill, CT. https://www.yelp.com/biz_photos/dinosaur-state-park-rocky-hill?select=FnmIO-BS23t2qep3fG0DQQ

Dilophosaurus. http://dinosaurpictures.org/Dilophosaurus-pictures

Dilophosaurus. mentalfloss.com/article/60231/10-crested-facts-about-dilophosaurus

Dinosaur state park, Rocky Hill, Connecticut. http://www.dinosaurstatepark.org/

5. Beneski Natural History Museum, Amherst College

Beneski Natural History Museum. https://www.amherst.edu/museums/naturalhistory

Noah's Raven trackway. https://www.dinotracksdiscovery.org/stories/noahs-raven/1/

Noah's raven trackway. http://www.nashdinosaurtracks.com/first-dinosaur-tracks.php

6. Summit of Sugarloaf Mountain

Pocumtuck Range. http://pocumtuck.org/

View from summit. http://elevation.maplogs.com/poi/deerfield_ma_usa.23704.html#

Map of Sugarloaf Mountain. www.franklinsites.*com/hikephotos/Massachusetts/southsugarloaf.php*

Map of Lake Hitchcock. http instaar.colorado.edumeetingsAW2013travel_infoamherst.html

Alluvial fan in Death Valley. http://www.newgeology.us/presentation13.html

Bajada of alluvial fans in Death Valley. http://www.earthonlinemedia.com/ebooks/tpe_3e/fluvial_systems/fluvial_processes_in_dry_regions.htm

Cross section of Deerfield basin. http://planetward.org/westernmassgallery.html

8. Poet's Seat Tower, Greenfield

Map of Poet's Seat Tower. http://www.franklinsites.com/hikephotos/Massachusetts/rockymtn-2006-0906.php

Poet's Seat Tower. http://www.greenfieldrecreation.com/parks.html

Poet's Seat Tower. https://newengland.com/yankee-magazine/living/profiles/poetry/

Poet's Seat Tower. https://www.britannica.com/place/Greenfield-Massachusetts

Poet's Seat Tower. http://www.atlasobscura.com/places/poet-s-seat-tower

10. Fluvial redbeds, Country Club Road, Greenfield

River avulsion. http://www.geo.uu.nl/fg/palaeogeography/results/avulsions

Formation of sole marks. http://www.columbia.edu/dlc/cup/ricci/ricci11.html

Grain lineation. http://strata.uga.edu/4500/xstrat2/xstrat2.html

Origin of cross-beds. www.kylefromohioblogspot.com

Beetle larvae in burrow. https://www.arkive.org/northern-dune-tiger-beetle/cicindela-hybrida/image-A4269.html

Origin of cross-beds. https://www.geocaching.com/geocache/GCZDCT_geology-along-the-trails-west-of-ncar?guid=b3885502-5a1f-4fe1-96d6-9551764dad0f

River with three-dimensional dunes. http://www.seddepseq.co.uk/SEDIMENTOLOGY/Sedimentology_Features/Ripples/Ripples.htm

Braided river. http://pages.uoregon.edu/millerm/braided.html

Susquehanna River flood. http://www.nytimes.com/2011/09/10/nyregion/ny-region-in-triage-mode-as-flooding-persists.html

11. Alluvial fan, West Gill Road, Gill

Origin of ventifacts. http://geologylearn.blogspot.com/2016/01/weathering-and-erosional-processes-in.html

12. Lacustrine strata, Barton Cove Campground, Gill

Map of Barton Cove. http://northampton.chambermaster.com/list/member/barton-cove-campground-northfield-4185.htm

Precipitation of calcite from lake. http://climatica.org.uk/wp-content/uploads/2014/03/21.png

Stratified meromictic lake. https://www.slideshare.net/GeromeRosario/lake-ecology-2017

Deerfield basin virtual field trip. http://minerva.union.edu/hollochk/keck_2008/index.html

Anticline and syncline. https://courses.lumenlearning.com/geology/chapter/reading-folds-anticlines/

Dexter Marsh. https://jurassicroadshow.com/2017/08/22/dexter-marsh/

Plunge pools and breccia. planetward.org/westernmassgallery.html

13. Lacustrine delta and alluvial-fan conglomerate, Chard Pond, Sunderland

Lake Manyara, Tanzania. http://www.alltimesafaris.com/destinations-tanzania-lakemanyara-national-park.html

14. Mount Toby Conglomerate, Roaring Brook, Sunderland

Roaring Brook. http://blogs.umass.edu/bikehara/2009/11/01/hiking-mount-toby/

Imbrication of pebbles. http://all-geo.org/highlyallochthonous/2008/02/imbrication-and-potholes-in-the-zebra-river/

Imbrication of pebbles. http://www.gigapan.com/gigapans?query=mount+toby

Imbrication of pebbles. https://www.slideshare.net/wwlittle/alluvial-fan-systems

Sam Lovejoy. http://www.gmpfilms.com/LNW.html

Sam Lovejoy. http://zpravodajstvi.ecn.cz/PRIVATE/Piano/Lovejoy.htm

Sam Lovejoy. http://www.masslive.com/news/index.ssf/2015/02/pipeline_foes_to_screen_film_a.html

Deposit of alluvial-fan flash flood. https://www.youtube.com/watch?v=FYffHjiFFlM

15. Deerfield Basalt, Route 2, Gill

Deerfield basin virtual field trip. http://minerva.union.edu/hollochk/keck_2008/index.html

Fissure eruption in Iceland. https://www.thetimes.co.uk/article/britain-faces-climate-chaos-from-toxic-icelandic-volcano-blast-hpptjljnpb6

Gases streaming from lava. http://www.abc.net.au/radionational/programs/scienceshow/lava-wing.jpg/5769684

Titled pipe vesicles. https://www.researchgate.net/publication/275968184_Vesicular_Basalts_as_a_Niche_for_Microbial_Life/figures?lo=1

16. Sugarloaf Arkose and hydrothermal hot spot, Greenfield

Photos of rock outcrops in Deerfield basin. http://planetward.org/westernmassgallery.html

Under-sea pillow lavas in Hawaii. http://khon2.com/2014/05/18/new-discovery-finds-oahu-is-made-up-of-three-not-two-volcanoes/

Cross-section of Deerfield basin. https://www.google.com/search?q=origin+of+alluvial+fan&rlz=1C1CHZL_en

Normal fault. https://www.mbmg.mtech.edu/kids/glossary.asp

Laurentide ice sheet. https://block6sciencetour.wordpress.com/things-to-know/

17. Geo Path and Rock Park, Greenfield Community College

Armored mudballs in river. http://ars.els-cdn.com/content/image/1-s2.0-S2095383615300936-gr8.jpg

Origin of armored mud balls. http://earthview.rocks/mudballs.html

Geo Path and Rock Park at Greenfield Community College. http://www.gcc.mass.edu/science/rock-park/

Armored mudballs. http://earthview.rocks/mudballs.html.

Glossary

Albite. Commonly white, albite is $NaAlSi_3O_8$, the sodium end member of the Na-Ca plagioclase solid solution series.

Alluvial fan. An alluvial fan is a cone-shaped deposit of sediment built up by intermittent (semi-arid) or continuous (humid) streams flowing from a mountain canyon onto a river plain or a playa. The apex of the fan is where the cone meets the mountain canyon. Deposition on the fan occurs because of the decrease in gradient at the break in slope between the mountain front and the valley floor or playa.

Analcime. White, gray, or colorless analcime is a hydrated sodium aluminum silicate with the composition $NaAlSi_2O_6 \cdot H_2O$. Some of the lacustrine strata in the Hartford Basin contain traces of analcime.

Anticline. A fold where each half (limb) of the fold dips away from the crest axis).

Antidune. In fast, shallow flows, antidunes form beneath surface standing waves of water. The bottom sand wave and the surface water wave are in phase. Antidunes can migrate up stream in contrast to dunes that migrate down stream. Antidune cross-beds dip upstream at low angles of about 10 degrees.

Avulsion. A river-channel avulsion occurs when a new channel is created by a rapid lateral shift of the channel to a new steeper path. The river flow in the new channel commonly erodes down into the underlying sediments.

Basalt. A gray to black, fine-grained igneous rock mainly composed of Ca-rich plagioclase and iron-bearing minerals such augite and olivine. The crystals are too small to see easily. Commonly there is a very fine-grained or glassy matrix interspersed with the mineral grains.

Bioturbation. When animals and plants rework sediments by burrowing and ingestion, they disturb the layering and texture of the deposit. At some locations in the Sugarloaf Arkose, the cross-beds and other primary sedimentary structures are completely lost.

Chlorite. The chlorite group of minerals has a layer-lattice, silicate structure. The four end members vary in their chemistry by substitution of Mg, Fe, Ni, and Mn in the silicate lattice. At location 15 behind the Super Stop and Shop store in Greenfield, the growth of minute crystals of a distinctive variety of chlorite in the sandstones shows that the rising hot water was between 150 and 200 degrees centigrade.

Continental crust. The 30 to 45 km-thick layer of igneous, sedimentary, and metamorphic rocks that forms the continents and continental shelves. The continental crust is richer in Si and Al and thus lighter in color and lower in density than the oceanic crust, which is darker and contains more Fe and Mg.

Dike. An intrusive, tabular igneous rock, the thickness of a dike varies from centimeter-scale to many meters, and the length can extend over many kilometers. A dike cross-cuts pre-existing layers or bodies of rock.

Dolomite. The calcium magnesium carbonate, dolomite $CaMg(CO_3)_2$. formed in some of the Triassic-Jurassic playa and lacustrine sediments by post-depositional alteration of Mg-calcite. Usually has a small amount of iron substitutes for magnesium, which causes the dolomite to weather yellow-brown.

Flute cast or mark. This feature is an erosional scour dug into unconsolidated sediment and then filled by the sand of the overlying bed (thus a cast). The long axis of the flute cast is the direction of water flow, with the steep end pointing up current.

Foliation. The sheet-like planar structure of foliation is present in some metamorphic rocks. The layers can vary in thickness from a sheet of paper to more than a meter. Foliation is based on the Latin word folium that means leaf.

Formation. A geological formation is a defined amount of strata that can be mapped. A formation can be subdivided into members. The Sugarloaf Arkose can be followed from place to place and mapped in the Deerfield basin.

Geothermal gradient. Away from tectonic plate boundaries, the rate of increase in the Earth's near surface temperature with increasing depth is about 25 to 30 °C/km (72 to 87 °F/mi). In rift valleys the gradient is commonly 10 degrees higher due to thinning of the Earth's crust.

Grain size. The grain-size scale is boulder (>256 mm), cobble (64-256 mm), sand (.062-64 mm), silt (.004-.062 mm) and clay (<.004 mm). Conglomerate comprises cobbles and boulders, whereas mud is silt and clay.

Granite. With 20 to 60% quartz and at least 35% potassium feldspar, granite is a common igneous rock. The granitic liquid cools slowly in the Earth's crust, resulting in easily visible crystals of quartz and feldspar. Granites are mostly white, pink, or gray depending on their mineralogy.

Half graben. A rift valley with a fault on one side. The Deerfield and Hartford basins are half grabens with the border fault on the east side.

Hematite. An iron-oxide mineral (Fe_2O_3) that is brown to reddish brown, or red, hematite is the main mineral that colors the Triassic-Jurassic redbeds of the Connecticut Valley.

Igneous. The three types of rocks are igneous, sedimentary, and metamorphic. Igneous rock forms by cooling and crystallization of intrusive magma or extrusive lava.

Imbrication. A river current can flip over disc-shaped pebbles and cobbles to produce a preferred orientation where the stones overlap one another in a consistent fashion, like toppled dominoes. The clasts dip in up flow direction.

Lava. At temperatures from 700 to 1,200 °C (1,292 to 2,192 °F), lava is hot, liquid rock erupted from a fissure or volcano.

Playa. Although playa is Spanish for a beach, in geomorphology a playa is a desert basin with no outlet that during a storm partially fills with water to form a temporary lake.

Planar cross-bed set. A set of cross-beds **is** planar when the surfaces that bound the set are planar rather than curved. Planar cross-bed sets form when the wave form made of sand has a straight crest and moves down current depositing cross-beds.

Precession cycle. The Earth's axis wobbles due to the pull of the Moon and Sun. For the axis to sweep out one complete cycle takes 26,000 years, called the precession cycle. Over time the precession cycle slowly lengthens so that back in the early Mesozoic it was 20,000 years.

Bajada. A bajada consists of a series of coalescing alluvial fans along a mountain front. The Sugarloaf Arkose at Sugarloaf Mountain is part of a bajada more than 625 km^2 in size.

Mg-calcite. Calcite is the calcium carbonate, $CaCO_3$. In Mg-calcite, magnesium substitutes for Ca, mostly as a few mole percent. Mg-calcite was a common precipitate in the early Mesozoic lakes of western Massachusetts because the lake waters had substantial Mg in solution.

Rift valley. An elongated rift valley forms by the depression of a block of the Earth's crust between two subparallel normal faults created by regional extension.

Schist. Characterized by platy grains of mica in parallel orientation, schist forms during medium-to-high grade regional metamorphism. Schist has more than 50% mica, commonly interlayered with quartz and feldspar.

Semiarid. A climate with low annual rainfall from 10 to 20 inches.

Strata. Strata are layers of sedimentary rock where each layer was deposited on top of the previous one. In some cases such as the sea floor, strata can extend over hundreds of thousands of square kilometers.

Syncline. A fold in which each half (limb) of the fold dips toward the trough (axis) of the fold.

Syndepositional. A sedimentary structure, such as the pillows at Chard Pond (location 12), are syndepositional when they form while deposition of the sand was proceeding.

Thin section. A thin slice of rock is ground flat and mounted on a glass slide. The rock is ground smooth using progressively finer abrasive grit until 30 microns thick. The slide is studied with a polarizing petrographic microscope or electron microscope to observe the minerals and texture of the rock.

Trough cross-bed set. Subaqueous dunes of sand with crestlines characterized by repeated curved recesses (cuspate) have scooped-out pits in front of the recesses. The dune advances down current depositing cross-beds of sand. The cross-beds fill each pit, producing a curved bounding surface to the cross-bed set.

Type section. The type section is the locality where a formation is first identified. Sugarloaf Mountain (locations 6 and 7) is the type section for the Sugarloaf Arkose.

Ventifact. A stone where wind-driven sand or silt has sculpted more or less flat surfaces (facets). Additional features are pits, groves, and polish. The ridge (keel) between the facets is parallel to the dominant direction of the wind.

www.ingramcontent.com/pod-product-compliance
Lightning Source LLC
Chambersburg PA
CBHW082329220526
45470CB00008B/2447